CARPENTRY
SOME TRICKS OF THE TRADE

BOB SYVANEN

The
Globe
Pequot
Press

OLD SAYBROOK, CONNECTICUT

MANUFACTURED IN THE UNITED STATES OF AMERICA
SECOND EDITION/FOURTH PRINTING

LIBRARY OF CONGRESS CATALOGING-IN-PUBLICATION DATA

SYVANEN, BOB, 1928–
    CARPENTRY, SOME TRICKS OF THE TRADE.

    1. HOUSE CONSTRUCTION.   2. CARPENTRY.   I. TITLE.
    TH4811.S97 1987   694'.2   87-32004
    ISBN 0-87106-783-8 (PBK.)

INTRODUCTION

DURING MY 30 YEARS AS AN ARCHITECT I HAVE DEALT WITH HUNDREDS OF CONTRACTORS, BUILDERS, CARPENTERS, AND CABINETMAKERS. MANY HAVE DONE BETTER THAN AVERAGE WORK. A FEW HAVE BEEN OUTSTANDING. SEEING VARIATIONS, I'VE OFTEN WONDERED WHAT IT WAS THAT MADE ONE PERSON'S WORK SO MUCH BETTER THAN ANOTHER'S. IT DIDN'T SEEM TO DEPEND ON INTELLIGENCE, EXACTLY, OR EXPERIENCE, OR EVEN MANUAL SKILLS. NOT COMPLETELY. THERE WAS SOMETHING ELSE INVOLVED. BUT IT WASN'T UNTIL I WATCHED BOB SYVANEN BUILD MY HOUSE THAT THE ANSWER BEGAN TO DAWN UPON ME.

THERE IS A DIRECT RELATIONSHIP BETWEEN GOOD CARPENTRY AND EFFICIENT WORK HABITS. SKILLED CARPENTERS SEEM TO PRODUCE LESS WASTE. THEY WASTE LESS TIME, LESS EFFORT, AND FEWER MATERIALS. THEY TURN MISTAKES TO THEIR ADVANTAGE. THEY KNOW THE TRICKS OF THE TRADE.

AFTER I'D SEEN BOB IN ACTION FOR SEVERAL MONTHS I ASKED HIM IF HE'D CONSIDER DOING A BOOK ON HIS CRAFT, NOT ON CARPENTRY AS SUCH—THE MARKET IS FLOODED WITH GOOD BOOKS ON THAT SUBJECT—BUT ON THE LITTLE SECRETS THAT SPELL THE DIFFERENCE BETWEEN CRISIS-TO-CRISIS CARPENTRY AND UNDER-CONTROL CARPENTRY. AS SOON AS BOB AGREED I TALKED HIM INTO GIVING ME THE ILLUSTRATING JOB.

IT MAY SEEM SLIGHTLY INAPPROPRIATE TO SEE SO MANY PAGES DEVOTED TO ORDINARY CARPENTRY WHEN THE WORLD IS RUNNING OUT OF FUEL, WHEN NEWER, MORE EFFICIENT, KINDS OF BUILDINGS ARE SO BADLY NEEDED. YOU MIGHT THINK WE SHOULD BE OFFERING INFORMATION INSTEAD ON, SAY, SOLAR HOUSE CONSTRUCTION. BUT THE MARKET IS FLOODED WITH GOOD BOOKS ON THAT SUBJECT TOO.

NO MATTER WHAT WE BUILD WE MUST LEARN TO BUILD IT EFFICIENTLY, TO DO IT RIGHT IN SPITE OF ALL THE JUNK THAT PASSES FOR BUILDING MATERIALS TODAY. THAT'S WHAT BOB'S BOOK IS ALL ABOUT.

MALCOLM WELLS
ARCHITECT
BREWSTER, MASSACHUSETTS

4

THE OLDER I GET THE MORE I AM CONVINCED THAT ANYONE CAN DO ANYTHING. ALL WE NEED IS ENOUGH DESIRE.

IN ANY ENDEAVOR THERE ARE THE LITTLE UNFORSEEN PROBLEMS THAT CAN CAUSE US TO STUMBLE, AND ALTHOUGH THERE ARE MANY "HOW TO" BOOKS ON ALL MANNER OF SUBJECTS THESE LITTLE UNFORSEEN PROBLEMS ARE RARELY MENTIONED. HOW WE DEAL WITH THESE PROBLEMS IS FREQUENTLY WHAT MAKES FOR A SUCCESS OR FAILURE IN OUR EFFORTS.

WE READ ABOUT JOISTS AND RAFTERS. WE LEARN WHERE THEY GO, HOW MANY NAILS, BUT WHAT SHALL I DO WHEN THE MEMBER IS WARPED AND DOESN'T DO WHAT IT IS SUPPOSED TO DO? IS THERE AN EASY WAY TO PICK UP LUMBER AND WALK WITH IT? MISTAKES OCCUR, AND ALTHOUGH THEY ARE NOT IN THEMSELVES BAD THEY DO CAUSE US TO STUMBLE.

MISTAKES ARE JUST A DIFFERENT WAY OF DOING SOMETHING - AND WITH A DIFFERENT CONSEQUENCE. MANY TIMES A SO-CALLED MISTAKE TURNS OUT TO BE THE EFFORT WITH THE BETTER CONSEQUENCE. THESE "MISTAKES" (OF WHICH I HAVE MADE QUITE A FEW) ARE GREAT LEARNING TOOLS. HOW WE DEAL WITH THEM IS WHAT SEPARATES THE GOOD FROM THE MEDIOCRE. WHEN A PROBLEM OCCURS, STAY LOOSE; THE SOLUTION IS ALWAYS THERE.

I GOT INTO CARPENTRY 30 YEARS AGO BECAUSE I WANTED TO BE A GOOD ARCHITECT. I WANTED TO KNOW ABOUT THE LITTLE PROBLEMS AND THEIR SOLUTIONS. SO FOR 30 YEARS I HAVE BEEN LEARNING TO BE A CARPENTER AND ARCHITECT, AND MY HOPE FOR THIS BOOK IS THAT IT WILL MAKE AT LEAST ONE JOB EASIER FOR SOMEONE. REMEMBER TWO THINGS:

MEASURE TWICE, CUT ONCE, AND
YOU ARE BUILDING A HOUSE, NOT A PIANO.

BOB SYVANEN
BREWSTER, MASSACHUSETTS

CONTENTS

LUMBER SPECIES

SPECIES OF LUMBER CAN ALSO CAUSE PROBLEMS, BUT WE CAN OVERCOME. THERE ARE 5 BASIC CHOICES:
1. DOUGLAS FIR
2. SOUTHERN OR YELLOW PINE
3. SPRUCE
4. HEMLOCK
5. NATIVE PINE (NEW ENGLAND)

DOUGLAS FIR

NUMBER ONE CHOICE, ALONG WITH YELLOW PINE, FOR FRAMING. WHEN I FIRST STARTED, FIR IS ALL I EVER SAW. IT IS STRONG, STRAIGHT, HEAVY, AND HARD. IT IS HIGHLY DECAY RESISTANT, MAKING IT IDEAL FOR FOUNDATION SILLS AND DECKS. THE PROBLEMS WITH FIR COME FROM THE FACT THAT IT IS HEAVY AND HARD. ON SOME OLD CHARTS FIR WAS LISTED AS A HARDWOOD. THERE ARE 5 THINGS TO BE AWARE OF WHEN WORKING WITH FIR:
1. HEAVY.
2. HARD TO CUT WITH A HAND SAW, PARTICULARLY IF SAW IS DULL.
3. HARD TO NAIL.
4. SPLITS EASILY.
5. BLEEDING PITCH POCKETS.
SOLUTION:

IT'S HEAVY SO GET STRONG.

THE HAND SAW IS NOT USED AS MUCH AS IN THE PAST SO THAT PRESENTS LESS OF A PROBLEM. HOWEVER THERE ARE TIMES WHEN IT MUST BE USED. MAKE SURE IT IS SHARP, AND CUT WITH A SLOW STEADY STROKE. BE SURE THE STOCK IS WELL SUPPORTED AND STEADY.

GOOD NAILING JUST TAKES PRACTICE AND A STRONG FOREARM. BEFORE I STARTED MY FIRST JOB, I USED TO PRACTICE NAILING EVERY DAY. I ALSO SQUEEZED A RUBBER BALL TO IMPROVE MY GRIP.

AVOID NAILING TOO CLOSE TO THE END OF A BOARD. DON'T USE A 16d NAIL WHEN AN 8d WILL DO.

BLEEDING PITCH POCKETS ARE HARD TO CONTROL. THEY POSE NO PROBLEM IF THE LUMBER IS USED AS A STUD OR JOIST, BUT IF USED AS AN EXPOSED BEAM THERE COULD BE TROUBLE. LOOK FOR A BROWN SPOT OR STREAK, ESPECIALLY A CRACK WITH THE TELL-TALE BROWN. USE IT WITH THE BLEMISH TURNED UP TO KEEP IT FROM DRIPPING PITCH ON WHATEVER IS BELOW. SOMETIMES SUCH SPOTS CAN BE CUT OUT AND PATCHED, BUT BEWARE: THE POCKET CAN BE QUITE EXTENSIVE.

SOUTHERN OR YELLOW PINE

HANDLES LIKE FIR BUT WITHOUT THE PITCH POCKETS. IT TOO IS HEAVY, STRAIGHT, AND STRONG. IT HAS A LOT OF RESIN THAT BLEEDS THROUGH PAINT.
SOLUTION:

SAME AS FOR FIR. TRY DIFFERENT SEALERS ON SAMPLES BEFORE PAINTING.

SPRUCE

SPRUCE SEEMS TO BE THE MOST COMMON FRAMING LUMBER ON THE EAST COAST. IT IS VERY LIGHT (PARTICULARLY WHEN DRY), EASY TO CUT, EASY TO NAIL, SOUNDS WONDERFUL. IT DOES HAVE 3 WEAK POINTS:

1. LOW DECAY RESISTANCE.
2. POOR NAIL HOLDING.
3. WEAK

SOLUTION:

AVOID USING WHERE MOISTURE IS A PROBLEM SUCH AS SILLS AND OUTDOOR DECKS. IF IT MUST BE USED WHERE MOISTURE IS PRESENT, PAINT, PROTECT WITH OVERHANGING OR CAPPING SHEETS OF METAL. USE PRESSURE TREATED FOR FOUNDATION SILLS.

USE ADEQUATE NAILING, PREFERABLY HOT DIPPED GALVANIZED NAILS.

USE PROPER SIZED TIMBER FOR DESIGN LOADS ON BUILDING.

HEMLOCK

HEMLOCK AND SPRUCE SEEM TO BE WHAT WE ARE OFFERED THESE DAYS, IN NEW ENGLAND, FOR FRAMING. HEMLOCK, WHEN DRY, IS QUITE HARD AND TENDS TO SPLIT. IT ALSO TWISTS BADLY. GREAT IF YOU ARE MAKING SKIS. HEMLOCK HAS ONE GOOD QUALITY THAT I KNOW OF. IT IS A GOOD MATCH FOR BIRCH AND CAN BE USED AS TRIM WITH BIRCH CABINETS AND DOORS. IT STAINS SIMILAR TO BIRCH. TEST A FEW SAMPLES FIRST. HEMLOCK CAN BE RECOGNIZED BY THE NATURAL BLUE STREAKS IN THE WOOD.

SOLUTION:

USE WHEN GREEN, AND NAIL IT QUICK BEFORE IT GETS AWAY.

DO NOT USE FOR OUTDOOR DECKS.

NATIVE PINE

NATIVE PINE OF NEW ENGLAND IS GOOD TO WORK WITH, BUT GETTING SCARCE. ONLY THE SMALL MILLS CUT IT. IT, TOO IS HIGHLY DECAY RESISTANT.

LUMBER FINISHES AND SEASONING

THE DIFFERENCES IN THE SPECIES AND CONDITION OF YOUR LUMBER CAN CAUSE PROBLEMS. THERE ARE 4 CONDITION-CHOICES:

1. AIR DRY OR KILN DRY, DRESSED - BOTH HAVE THE SAME MOISTURE CONTENT AT THE END OF THEIR DRYING PROCESS. AIR DRY TAKES 4 MONTHS TO A YEAR, DEPENDING ON SIZE. KILN DRY TAKES 1/4 THE TIME. BOTH ARE PLANED SMOOTH. THE MARK ON THE LUMBER WILL BE "S-DRY" OR "KILN DRY".

2. GREEN DRESSED - FRESHLY CUT AND PLANED SMOOTH, FULL OF MOISTURE. THE MARK WILL BE "S-GREEN".

3. GREEN ROUGH CUT - FRESHLY CUT. ROUGH EXTERIOR, FULL OF MOISTURE. NO MARK, THE SURFACE IS TOO ROUGH TO TAKE A STAMP.

4. DRY ROUGH CUT - ROUGH EXTERIOR, USUALLY AIR DRY. NO MARKS, TOO ROUGH. IT IS VERY OBVIOUS, BY THE WEIGHT OF THE LUMBER, TO TELL IF IT IS DRY.

SOME OTHER MARKS YOU WILL SEE ARE, "D" FOR DRY, M.C. 15% FOR MOISTURE CONTENT 15%.

AIR DRY OR KILN DRY, DRESSED

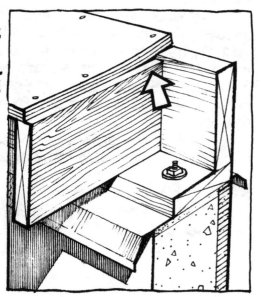

THIS IS THE EASIEST TO WORK WITH. MIXING AIR DRY AND KILN DRY CAUSES PROBLEMS IF THE AIR DRY HAS BEEN DRYING FOR TOO SHORT A TIME. THE DIMENSIONS OF THIS LUMBER WILL BE LARGER THAN THE KILN DRIED AND IF YOU USE THEM TOGETHER, PROBLEMS.

SOLUTION:

USE ONLY THE KILN DRIED OR AIR DRIED WHEN USING DIFFERING DIMENSIONED LUMBER YOU ARE CONSTANTLY COMPENSATING BY SHIMMING UP OR TRIMMING OFF (AND SOMETIMES FORGETTING). IF USING BOTH, DO YOUR COMPENSATING ACT, AND USE THE MATERIAL WHERE IT WILL MATTER THE LEAST, LIKE RAFTERS.

GREEN DRESSED

FOUR PROBLEMS HERE:

1. HEAVY
2. LUMBER WILL SHRINK
3. ROTTING AND FUNGUS ARE ENCOURAGED.
4. DOES NOT SHOW PENCIL MARK WELL.

SOLUTION:

THE WEIGHT OF GREEN LUMBER CAN BE OVERCOME. GO ON A GOOD WEIGHT LIFTING PROGRAM. ONE MAN CAN LIFT ONE 2X10X16' GREEN PLANK, TWO CAN BE HANDLED DRY.

YEARS AGO THE FRAMED HOUSE DRIED IN THE SUMMER SUN. IT CAN STILL BE DONE THAT WAY OR A FEW PORTABLE HEATERS CAN BE USED. PRACTICE GOOD FIRE PREVENTIVE METHODS. PROVIDE VENTILATION TO REMOVE MOISTURE FROM HOUSE AS

IT IS DRIVEN FROM THE LUMBER.

SEPARATE LAYERS OF LUMBER WITH STICKS AS SOON AS IT IS DELIVERED. USE 1" STICKS FOR THIN BOARDS AND 2" FOR THICK BOARDS.

USE AN INDELIBLE PENCIL FOR MARKING WET LUMBER, BUT BEWARE OF "BLUE TEMPLE". IF YOU CARRY THE PENCIL BEHIND THE EAR IT WILL LEAVE ITS MARK EACH TIME.

DRY ROUGH

THE BIG PROBLEM HERE IS THE UNEVEN DIMENSIONS. A 2x4 MAY MEASURE 2"x4" AT ONE END AND $2\frac{1}{3}$"x$2\frac{1}{4}$" AT THE OTHER. A 1x6 COULD BE $1\frac{1}{8}$" AND $\frac{7}{8}$". THE LARGER DIMENSIONS VARY NOT ONLY FROM END TO END, BUT ALSO FROM TIMBER TO TIMBER. ONE MAY MEASURE $9\frac{3}{4}$" ANOTHER $10\frac{1}{4}$". THE BETTER MILLS HAVE BETTER CONTROL, BUT NOT ABSOLUTE. THAT IS ONE OF THE REASONS FOR DRESSING LUMBER. IT IS EASIER TO CONTROL THE FINISHED SIZE WITH A PLANER.
SOLUTION:
BE AWARE OF VARIATIONS AND COMPENSATE BY SHIMMING AND TRIMMING. USE WHERE IT MATTERS LEAST.

GREEN ROUGH

GREEN ROUGH HAS THE SAME PROBLEMS AS DRY ROUGH WITH THE ADDED BONUS OF HEAVY AND MORE SHRINKAGE.
SOLUTION:
BE AWARE OF VARIATIONS AND START A REAL GOOD WEIGHT LIFTING PROGRAM. ALEXIEV OR ANDRE THE GIANT WOULD BE A GREAT ASSET.

11

MOVING LUMBER

ALL THESE METHODS WILL MOVE LUMBER, BUT NOT WITHOUT WASTING TIME AND ENERGY. NONE OF THESE SHOW GOOD CONTROL SO THEY ARE NOT SAFE. A MAN AT EACH END WILL MAKE EASY WORK OF THE JOB, BUT ALSO TWICE AS LONG.

TO PICK UP A PIECE OF LUMBER IS NOT A BIG PROBLEM, BUT FORTY 2X10'S ARE. THE BEST WAY TO CARRY LUMBER IS ON THE SHOULDER AND THE QUICKEST IS "THE LAZY MAN'S LOAD", TWO AT A TIME (TWENTY TRIPS INSTEAD OF FORTY). IF THE LUMBER IS GREEN, EVEN TWO MIGHT BE A BIT MUCH TO HANDLE. WHEN THE LOAD IS CARRIED ON THE SHOULDER, THE STICKS MUST POINT IN THE DIRECTION YOU ARE WALKING. WHEN CARRIED AT WAIST LEVEL, THE LOAD IS PERPENDICULAR TO THE DIRECTION YOU ARE WALKING AND THE WAY OF THE LEGS. IF TWO BOARDS ARE CARRIED THAT WAY THEY WILL SCISSOR.

TO PICK UP A LOAD PROPERLY, STAND AS CLOSE TO THE LOAD AS POSSIBLE WITH KNEES BENT AND BACK STRAIGHT. THE ARMS WILL HANG STRAIGHT AT THE ELBOWS, WITH THE LEFT HAND REACHING OVER TO GRAB THE FORWARD EDGE OF THE LOAD AND THE RIGHT HAND GRABBING THE REAR END OF THE LOAD, THUMBS UP. WITH THE ARMS STRAIGHT AND THE BACK KEPT FLAT, THE POWERFUL LEG MUSCLES DO THE LIFTING. DON'T JERK THE LOAD OFF THE GROUND OR THE BACK WILL BEND. INSTEAD DRIVE HARD WITH THE LEGS AND AS YOU PASS YOUR KNEES, SHRUG YOUR SHOULDERS HARD. WHEN THE LOAD IS AS HIGH AS IT WILL GO, DROP UNDER IT AND AT THE SAME TIME PIVOT THE BODY AND BOTH FEET TO THE LEFT 90° WHILE ROLLING THE LOAD TO CATCH ON THE SHOULDER. THE LUMBER FACE THAT WAS UP IS NOW DOWN. TO CUSHION THE LOAD AS IT LANDS ON THE SHOULDER, RAISE THE RIGHT ARM UP AND FORWARD. THE LOAD WILL THEN LAND ON THE SHOULDER MUSCLE INSTEAD OF THE BONE. BALANCE THE LOAD BY BOUNCING IT OFF THE SHOULDER AND QUICKLY SHIFTING UNDER IT. ONCE THE LOAD IS BALANCED, ONLY THE RIGHT ARM, RAISED AND STRETCHED FORWARD ON TOP OF THE LOAD, IS NEEDED FOR CONTROL AS YOU WALK.

MOVING LUMBER

BACK AND ARMS
STRAIGHT........

A SHRUG WITH
THE SHOULDERS.....

PIVOT, DIP, AND
ONTO THE RIGHT
SHOULDER WITH
RIGHT ARM RAISED....

THE LIFT IS COMPLETE ("THREE WHITE LIGHTS" AS THEY
SAY IN WEIGHTLIFTING) AND YOU ARE ON YOUR WAY.

THE LOAD CAN BE TOSSED FROM ONE SHOULDER TO THE OTHER BY THRUSTING WITH THE LEGS.
TO DROP THE LOAD EITHER ROLL IT OFF YOUR SHOULDER ONTO YOUR RIGHT FOREARM OR
TOSS IT UP AND AWAY WITH A LEG THRUST. PRACTICE RIGHT AND LEFT SHOULDER
WITH A FEW LIGHT BOARDS.

14

THIS DOESN'T
FEEL RIGHT.

MOVING SHEETS OF SHEETROCK OR PLYWOOD IS NOT ONE OF YOUR FUN JOBS, BUT THERE IS A BETTER WAY THAN SHOWN HERE.

DOESN'T THIS
DO SOMETHING
FOR MY POSTURE?

HEY GUYS,
CHECK THIS
OUT.

WHERE'S THE
DOOR?

HOW DID THIS
GET BEHIND
ME?

15

MOVING LUMBER

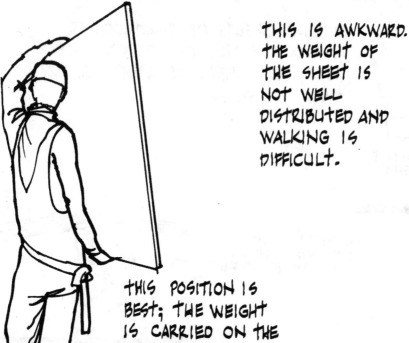

THIS IS AWKWARD. THE WEIGHT OF THE SHEET IS NOT WELL DISTRIBUTED AND WALKING IS DIFFICULT.

THIS POSITION IS BEST; THE WEIGHT IS CARRIED ON THE BOTTOM HAND. THE TOP HAND KEEPS THE SHEET FROM TIPPING.

IT'S A STRUGGLE WHEN TWO PEOPLE CARRY ONE SHEET USING OPPOSING GRIPS. EACH PERSON SHOULD HAVE THE SAME CARRYING HAND DOWN (RIGHT OR LEFT HAND IN THE BEST CARRYING POSITION).

HOW THE LUMBER IS STACKED AND WHERE IT SITS ON THE BUILDING SITE CAN BE A GREAT HELP IN MINIMIZING PROBLEMS. WHEN MAKING A LIST FOR THE YARD, INDICATE WHAT THE MATERIALS ARE FOR. BE SURE TO KEEP A COPY FOR YOURSELF. THE LUMBER YARD WILL THEN BE ABLE TO STACK THE LUMBER ON THE TRUCK IN A GOOD SEQUENCE. SOMETIMES YARDS DO A GOOD JOB AND SOMETIMES NOT, AT ANY RATE IT SHOULD BE STACKED WITH THE BUILDING SEQUENCE IN MIND.

LUMBER LIST		
EXTERIOR STUDS	2×6	400/8'
INTERIOR STUDS	2×4	420/8'
FIRST FLOOR JOISTS	2×10	50/14'
HEADERS	2×10	7/12' 10/14'
SECOND FLOOR JOISTS	2×10	50/14'

WHEN ORDERING LUMBER DON'T FORGET TO ORDER AT LEAST A HALF DOZEN EXTRA 2×10×14'-0" PLANKS FOR SCAFFOLDING. ALSO 2 OR 3 BUNDLES OF CHEAP WOOD SHINGLES FOR SHIMMING, 2 DOZEN 1×6×16'-0" FENCE BOARD (AN INEXPENSIVE SOUND BOARD FOR BRACING), AND 2 BUNDLES OF 1×3 FURRING STRIPS FOR BRACING.

BEFORE THE LUMBER TRUCK ARRIVES, THINK ABOUT WHAT PROBLEMS THE STACK OF LUMBER WILL CAUSE:

1. EASE OF ACCESS FOR TRUCK.
2. EASE OF UNLOADING.
3. RESTACKING OF LOAD.
4. MOVING LUMBER FROM STACK TO BUILDING.
5. LOCATION OF RADIAL SAW AND POWER POLE
6. MASONRY SUPPLIES.
7. TRAFFIC FLOW.

GREEN TIMBER SHOULD BE STACKED CRISSCROSSED AND THIN BOARDS STACKED WITH 1" SPACERS. ALIGN SPACERS VERTICALLY OR THIN BOARDS WILL DRY CURVED. KEEP ALL GOOD LUMBER OFF THE GROUND. IN THE ALPINE COUNTRIES, TIMBER STACKS WERE DRIED FOR 3 YEARS BEFORE USE. THE THIN BOARDS WERE STORED FOR 1 YEAR.

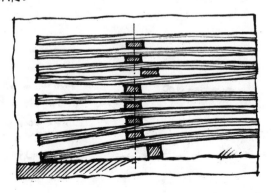

17

PLYWOOD CAN BE SECURED BY NAILING THE CORNERS DOWN WITH 16d NAILS, FOR HALF OR MORE OF THE STACK. IT HELPS KEEP THE STACK FROM WALKING AWAY. REMOVING NAILS IN THE DARK IS DIFFICULT AND NOISY. COVER WITH PLASTIC AND PLENTY OF WOOD STRIPS.

KILN DRY LUMBER SHOULD BE COVERED WITH A TARP; PLASTIC WORKS WELL. USE PLENTY OF WOOD STRIPS NAILED THROUGH THE PLASTIC TO THE STACK INSIDE. THIS COVER KEEPS THE RAIN, SNOW, AND SUN OFF AND MAKES IT DIFFICULT FOR THE LUMBER TO WALK OFF. I BUILT A PLYWOOD WALL AROUND A STACK OF LUMBER TO KEEP IT FROM DISAPPEARING.

PLOP PLOP PLOP....

STAND ROLLS OF BUILDING PAPER UNDER COVER ON THEIR ENDS ON A PIECE OF PLYWOOD. ROLLS THAT LAY ON THEIR SIDES GET OVAL SHAPED AND DON'T UNROLL TOO EASILY. IF THEY SIT ON THE GROUND THE ENDS PICK UP DIRT AND MOISTURE, AGAIN MAKING UNROLLING DIFFICULT.

WHEN USING TEMPORARY POSTS, CUT THEM ABOUT 3/4" SHORTER THAN THEY SHOULD BE. IT IS A SIMPLE JOB TO SHIM UP WITH WOOD SHINGLES. USE ANY JACKING DEVICE AVAILABLE TO RAISE THE BEAM; HYDRAULIC, SCREW, AUTO, EVEN A LEVER.

IT IS EASIER TO PUT THE POST ON THE FOOTING WITH THE JACK ON TOP. IF THE JACK IS ON THE BOTTOM, THE POST THEN HAS TO BE BALANCED ON THE JACK WITH ONE HAND WHILE YOU WORK THE JACK WITH THE OTHER. OF COURSE WITH TWO PEOPLE, EITHER WAY WILL WORK.

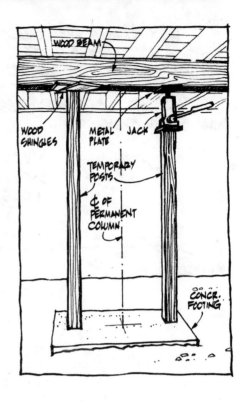

THE TEMPORARY POSTS CAN BE SECURED WITH A BRACE EACH WAY. ONCE FLOOR JOISTS ARE IN THE BRACES CAN BE REMOVED.

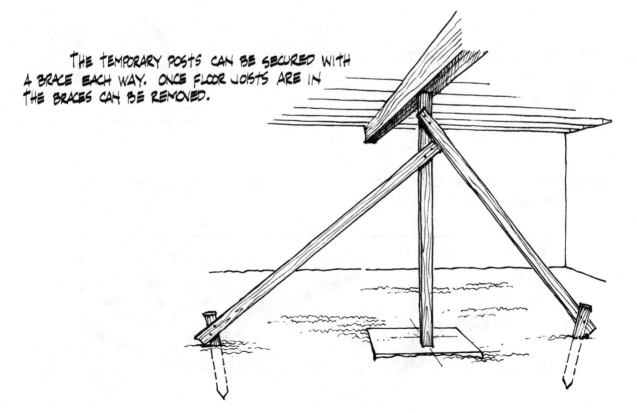

COLUMNS AND POSTS

LALLY COLUMNS ARE CONCRETE FILLED STEEL TUBES. THE EASIEST WAY TO CUT THEM IS WITH A LARGE TUBING CUTTER. MOST LUMBER YARDS HAVE THEM AND THEY WILL CUT THE COLUMNS, BUT YOU TAKE A CHANCE ON THEIR ACCURACY. THE BEST THING TO DO IS TO MEASURE THEM UP. DEDUCT FOR THE TOP AND BOTTOM PLATES. TAKE THEM TO THE YARD AND CUT THEM YOURSELF. IF THEY COME UP A LITTLE SHORT, METAL SHIMS WILL BRING THEM UP.

YOU CAN ALSO CUT THEM ON THE JOB WITH A HACKSAW. THE ENDS ARE ALWAYS SQUARE, SO JUST MARK THE LENGTH 4 OR 5 PLACES AROUND THE COLUMN.

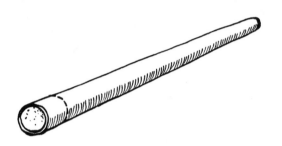

CUT MOST OF THE WAY THROUGH THE METAL CASE ALL AROUND.

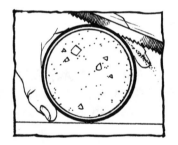

RAISE THE CUT END AND REST ON A BLOCK OR ANOTHER COLUMN.

SMACK WITH SLEDGE ········ PRESTO.

20

WHEN SHIMMING THE ENDS OF BEAMS RESTING ON CONCRETE OR STEEL, USE A NON COMPRESSIBLE MATERIAL LIKE STEEL, SLATE, OR ASBESTOS BOARD. ASBESTOS BOARD WORKS WELL BECAUSE IT COMES IN 1/8" SHEETS, WHICH MAKES IT EASY TO BUILD UP, AND IT CUTS EASILY.

SHIM
BELOW BEAM

THE SHIMS ARE TO KEEP THE BEAM UP PERMANENTLY AND IF WOOD IS USED IT WILL SHRINK, COMPRESS, AND MAYBE EVEN ROT AWAY. PLYWOOD WORKS PRETTY WELL; ITS COMPRESSION LOSS IS MINIMAL.

SILLS ARE THE MOST IMPORTANT STEP IN THE BUILDING. IF ALL IS SQUARE AND LEVEL THEN THE REST WILL GO QUITE EASY.

STEPS FOR SETTING SILLS

1. ESTABLISH TRUE AND SQUARE CORNERS.
2. ESTABLISH THE INSIDE CORNERS OF SILLS.
3. MARK INSIDE EDGE OF SILLS.
4. MATERIAL FOR SILLS.
5. LOCATE ANCHOR BOLT HOLES ON SILL, DRILL.
6. SETTING SILLS IN PLACE.
7. LEVELING SILLS.
8. INSTALL TERMITE BARRIER.

ESTABLISHING TRUE AND SQUARE CORNERS

AFTER THE FOUNDATION HAS BEEN COMPLETED CHECK ITS LENGTH AND WIDTH PER PLAN. IF SIDES ARE EQUAL AND DIAGONALS ARE EQUAL THEN THE FOUNDATION IS SQUARE. IF NOT THEN AN ADJUSTMENT MUST BE MADE.

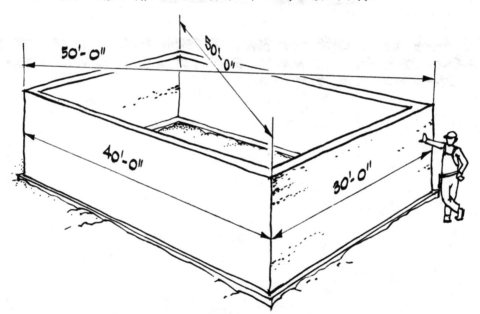

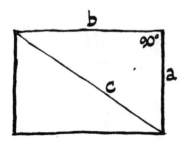

THE EXACT DIAGONAL CAN BE FOUND MATHEMATICALLY. $a^2 + b^2 = c^2$ AND WITH THE NEW SMALL POCKET CALCULATORS IT IS EASY. THERE ARE SPECIAL SLIDE RULES THAT GIVE THE DIAGONAL WITH A TWIST OF A DIAL. JUST KEEP ADJUSTING CORNERS UNTIL YOU HIT THE DIMENSIONS REQUIRED.

IF YOU HAVE A FOUNDATION THAT IS MORE THAN A SIMPLE RECTANGLE, THE 3, 4, 5 TRIANGLE IS VERY HELPFUL. AFTER THE MAIN FOUNDATION IS SQUARED AN ADDITION CAN BE SQUARED OFF ANY WALL WITH THE 3, 4, 5 TRIANGLE. THE 3, 4, 5, TRIANGLE IS THE SAME AS $a^2 + b^2 = c^2$ ($3^2 + 4^2 = 5^2$).

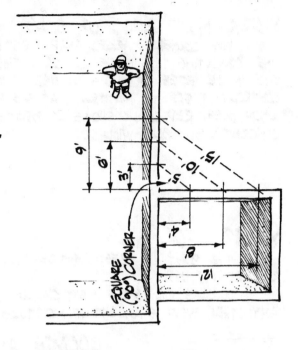

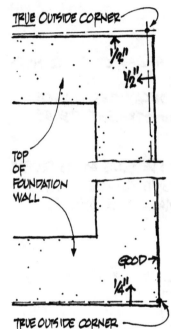

TRUE OUTSIDE CORNER

TOP OF FOUNDATION WALL

GOOD

TRUE OUTSIDE CORNER

WHEN THE CORNERS ARE ESTABLISHED, MARK THEM ON THE FOUNDATION. IF A CORNER IS ½" OUTSIDE OF THE CONCRETE, MARK ½" ↑. IF IT IS ½" IN ON THE CONCRETE, MARK ½" ↓. IF AN EDGE IS RIGHT ON, MARK "GOOD."

TRUE INSIDE CORNER

TAKE A SHORT BLOCK OF THE SILL MATERIAL AND PLACE IT AT THE CORNER, THE TRUE CORNER. MARK THE INSIDE FACE ON THE FOUNDATION. USE A NAIL TO SCRATCH A GOOD MARK EACH WAY FORMING AN + FOR THE TRUE INSIDE CORNER.

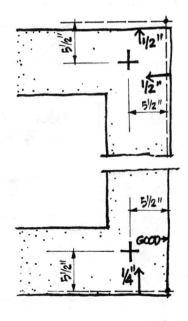

MARK INSIDE EDGE OF SILLS

WITH CORNERS MARKED, SNAP A RED CHALK LINE CONNECTING THE INSIDE CORNERS. THIS IS THE INSIDE EDGE OF THE SILL. USE IT FOR CONTROL; IT DOESN'T MATTER WHAT THE FOUNDATION DOES. RED CHALK SHOWS UP BETTER ON CONCRETE, BUT BLUE WILL DO.

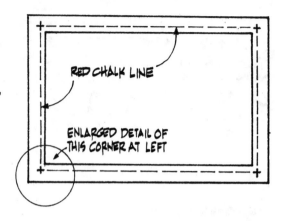

RED CHALK LINE

ENLARGED DETAIL OF THIS CORNER AT LEFT

MATERIAL FOR SILLS

THE BEST IS PRESSURE TREATED, AND IN FACT MANY AREAS OF THE COUNTRY REQUIRE ITS USE.

DOUGLAS FIR IS THE NEXT CHOICE BECAUSE OF ITS HIGH DECAY RESISTANCE. ANY OTHER WOOD IS ALRIGHT IF MOISTURE IS KEPT AWAY. CHOOSE STRAIGHT PIECES.

LOCATE ANCHOR BOLTS

REST THE SILL ON THE FOUNDATION ALONG SIDE BOLTS AT ITS PROPER LOCATION RELATIVE TO THE TRUE OUTSIDE CORNER. WITH A COMBINATION SQUARE, LOCATE AND MARK THE CENTER LINE OF BOLTS.

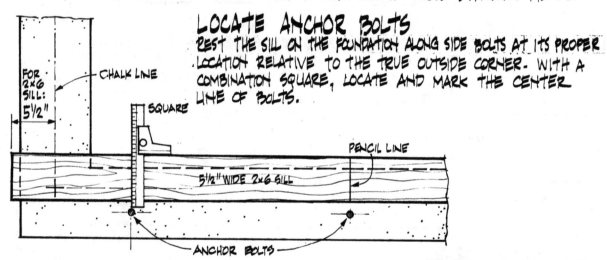

FOR 2×6 SILL: 5½"

CHALK LINE

SQUARE

PENCIL LINE

5½" WIDE 2×6 SILL

ANCHOR BOLTS

MEASURE FROM CENTER LINE OF BOLTS TO CHALK LINE, TRANSFER TO SILL. DRILL 7/8" HOLE FOR ½" BOLT TO ALLOW FOR ADJUSTING.

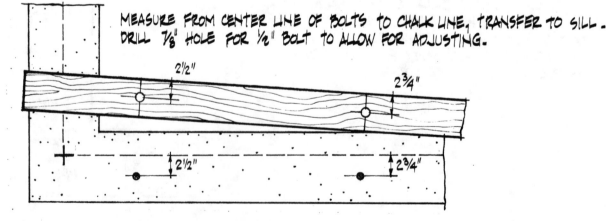

2½"

2¾"

2½"

2¾"

THERE ARE USUALLY
A FEW ANCHOR BOLTS
LEANING.......

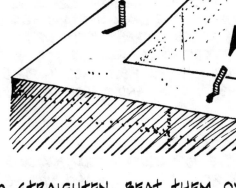

TO STRAIGHTEN, BEAT THEM OVER
WITH A HAND MAUL, BUT SCREW
A NUT ON AND HIT IT INSTEAD OF
THE BOLT THREADS.

LOCATING ANCHOR BOLT
HOLES IS EASY WITH THIS
TOOL. ALIGN THE OUT-
SIDE FACE OF THE SILL
WITH THE CHALKLINE.
WITH THE TOOL IN PLACE
HAMMERING ON THE SELF-
TAPPING SCREW LOCATES
THE BOLT HOLE.

5½"

5½"

5½"

5½"

EYEBALL 90°

CHALK LINE AT INSIDE
EDGE OF MUDSILL

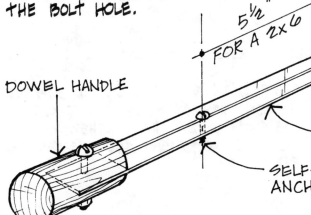

5½"
FOR A 2×6

ANCHOR BOLT

DOWEL HANDLE

BLADE FROM
COMBINATION SQUARE

SELF-TAPPING SCREW MARKS
ANCHOR BOLT HOLE LOCATION

SILLS

INSTALLING SILLS

LIGHTLY BOLT THE FIRST CORNER SILL PIECE IN PLACE LINING UP WITH THE CHALK LINE AND TRUE OUTSIDE CORNER. THE NEXT PIECE IS THEN BUTTED UP AGAINST IT AND THE SAME PROCEDURE IS FOLLOWED FOR BOLT LOCATIONS. WHEN ALL THE SILL PIECES ARE IN PLACE, RECHECK LENGTHS, WIDTHS, AND DIAGONALS. ADJUST IF NECESSARY.

LEVELING SILLS

THE MOST ACCURATE TOOL FOR LEVELING IS THE WATER TUBE LEVEL. IT IS JUST A LONG PIECE OF CLEAR PLASTIC TUBING WITH A COLORED ANTIFREEZE SOLUTION IN IT. THERE ARE GARDEN HOSE ATTACHMENTS AVAILABLE BUT THE PROBLEM WITH GARDEN HOSE IS THAT YOU CAN'T SEE IF THE LINE IS CLEAR OF BUBBLES. THE LINE MUST BE CLEAR OF BUBBLES OR YOU WILL NOT GET A TRUE READING.

USING THE TUBE IS VERY SIMPLE. KEEP ONE END AT THE HIGH POINT OF THE FOUNDATION AND MOVE THE OTHER TO VARIOUS POINTS AROUND THE WALL. THE WATER WILL SEEK THE SAME LEVEL AT EACH END AND YOU CAN NOTE THE VARIATIONS IN WALL HEIGHTS. MARK THESE HEIGHT DIFFERENCES ON THE WALL, $+\frac{1}{2}"$, $-\frac{1}{4}"$, ETC.

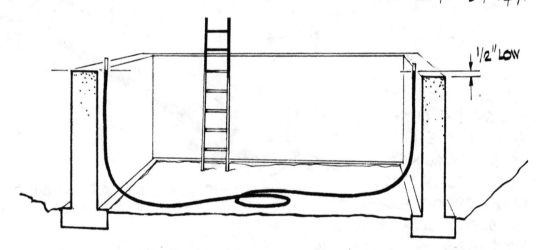

1/2" LOW

SIGHT LINE

LEVEL LINE

INSTRUMENT OUT OF LEVEL

THE BUILDER'S LEVEL IS GOOD AS LONG AS CARE IS TAKEN WHEN SETTING UP. TRY TO KEEP THE "SHOTS" SHORT. THE FURTHER AWAY FROM THE INSTRUMENT THE GREATER THE ERROR.

26

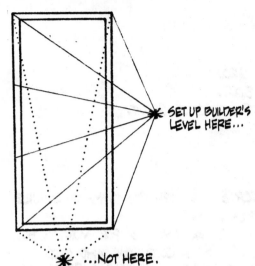

SET UP BUILDER'S LEVEL HERE...

...NOT HERE.

THE INSTRUMENT CAN BE SET UP INSIDE, AT THE CENTER, OF A LOW FOUNDATION WALL TO GOOD ADVANTAGE. WHEN SHOOTING FROM OUTSIDE, SET UP ALONG THE LONG WALL TO KEEP "SHOTS" SHORT AND SOMEWHAT EQUAL.

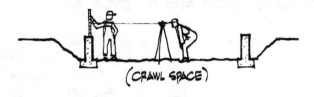

(CRAWL SPACE)

SET UP A REFERENCE POINT THAT YOU CAN CHECK EASILY. SOME ONE IS BOUND TO KICK ONE OF THE TRIPOD LEGS. YOU CAN USE A NAIL IN A TREE OR SOME STATIONARY OBJECT, A RULER, OR A WOOD SOLDIER.

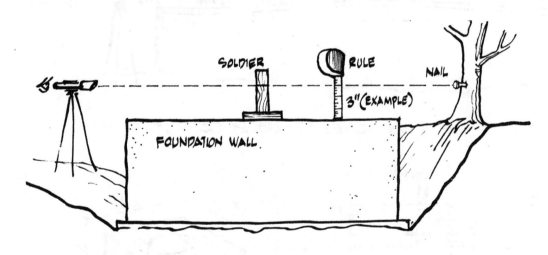

SOLDIER   RULE   NAIL   3" (EXAMPLE)   FOUNDATION WALL

AS LONG AS THE CROSS HAIRS HIT THE ORIGINAL REFERENCE MARK (NAIL IN TREE, MARK ON RULER, MARK ON WOOD SOLDIER) ALL IS WELL. IF THE INSTRUMENT IS MOVED A NEW REFERENCE POINT MUST BE ESTABLISHED. KEEP CHECKING, THERE IS AN OLD CARPENTER'S SAYING, "MEASURE TWICE CUT ONCE!"

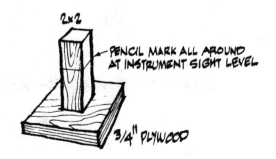

2x2

PENCIL MARK ALL AROUND AT INSTRUMENT SIGHT LEVEL

3/4" PLYWOOD

SILLS

STEPS IN LEVELING SILLS:
1. LOCATE HIGH POINT.
2. TAKE READING ON RULER.
3. TAKE READING AT EACH BOLT AND ADD ½" FOR GROUT.
4. CUT WOOD SHIM OF THICKNESS REQUIRED AT EACH BOLT.
5. PAINT CEMENT WASH ON FOUNDATION TOP TO AID BONDING OF GROUT.
6. PUT MORTAR BED IN PLACE.
7. PUT SILLS ON.
8. FINISH EXPOSED EDGES OF MORTAR BED.

1. WITH SILLS IN PLACE, LOCATE THE HIGH POINT, MAKING SURE SILL IS IN CONTACT WITH FOUNDATION.

2. & 3. TAKE A READING ON A RULER AND ½" FOR MORTAR BED. THE BED SERVES TWO PURPOSES: A GOOD SEAL AND GOOD SUPPORT. DO THE SAME AT EACH BOLT, MAKING SURE ½" IS ADDED. RECORD DIMENSIONS ON SILL.

3½"   3¾"   3"   3⅛"   3¾"   3⅜"   3½"

4. CUT SHIMS, 2"x2"x THICKNESS REQUIRED AT EACH BOLT.

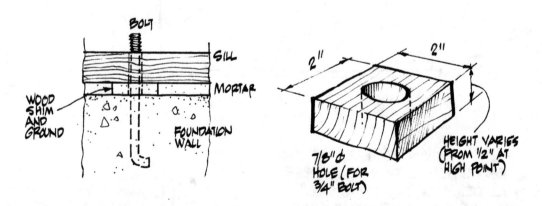

BOLT
SILL
MORTAR
WOOD SHIM AND GROUND
FOUNDATION WALL

2"   2"
7/8"⌀ HOLE (FOR ¾" BOLT)
HEIGHT VARIES (FROM ½" AT HIGH POINT)

5. PAINT TOP OF FOUNDATION WITH A THIN CREAMY PASTE OF CEMENT AND WATER.

6.&7. MIX A RICH BATCH OF MORTAR AND APPLY TO TOP OF FOUNDATION WALL. KEEP IT SOFT SO THAT THE SILL WILL HAVE NO TROUBLE COMPRESSING IT TO THE LEVEL OF THE SHIMS. KEEP THE MORTAR AWAY FROM SHIMS TO MAKE SURE IT DOES NOT OOZE OVER ON TOP PREVENTING THE SILL FROM SITTING ON THE SHIMS.

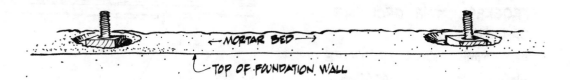

←—MORTAR BED—→

└—TOP OF FOUNDATION WALL

IF SILL PIECE IS WARPED, THAT IS, HAS A BOW IN IT, PUT THE HUMP UP AND PRESS IT INTO THE MORTAR WITH A STRONG STRAIGHT EDGE.

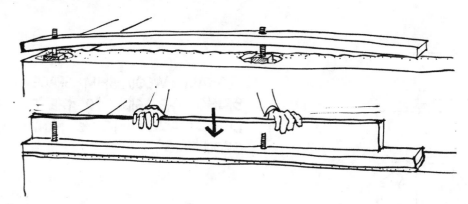

IF SILL PIECE HAS A "CROOK", SOMETIMES CALLED A "CROWN" OR "WOW", IT CAN BE USED IF YOU SNAP A CHALK LINE AND CUT IT STRAIGHT.

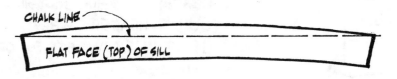

CHALK LINE

FLAT FACE (TOP) OF SILL

8. CLEAN UP SQUEEZED OUT MORTAR, AND JOINT WITH TROWEL.

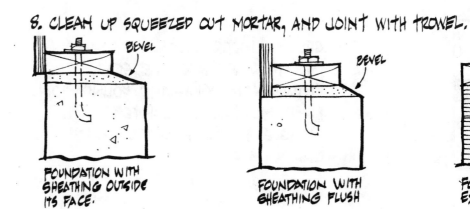

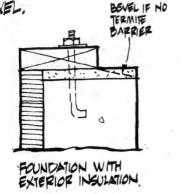

BEVEL IF NO TERMITE BARRIER

BEVEL

FOUNDATION WITH SHEATHING OUTSIDE ITS FACE.

BEVEL

FOUNDATION WITH SHEATHING FLUSH

FOUNDATION WITH EXTERIOR INSULATION.

29

GROUTING THE MUDSILL
AFTER IT'S INSTALLED
ALLOWS THE FRAMING TO
PROGRESS. THE GROUTING
CAN THEN BE DONE AT
ANY TIME. HOWEVER THERE
SHOULD BE EASY ACCESS
TO BOTH SIDES OF THE SILL.

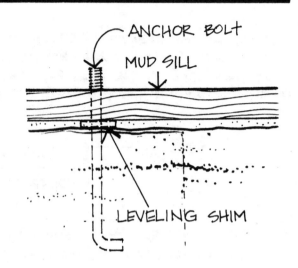

INSTALL WOOD SHIM SPACERS TO
BRING THE SILL TO THE PROPER
LEVEL (ALLOWING ½" MINIMUM
GROUTING).

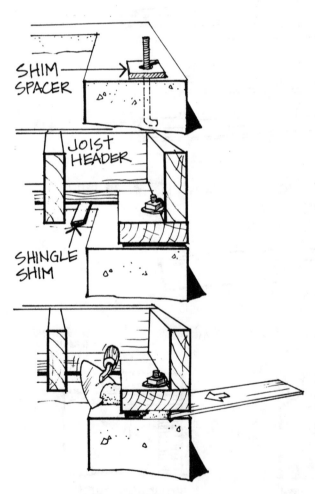

INSTALL MUDSILL, JOISTS, AND
HEADERS. SHIM MUDSILL UP TO
JOIST HEADER WHERE NECESSARY
USING NARROW STRIPS OF WOOD
SHINGLES.

GROUT UNDER MUDSILL BY THROW-
ING MORTAR INTO THE SPACE
BETWEEN MUDSILL AND FOUNDATION.
PUSH MORTAR TO CENTER WITH
SHINGLE BUTT. FILL ALL SPACE
UNDER MUDSILL.

TERMITE BARRIER

WHERE TERMITES ARE A PROBLEM, A BARRIER IS A MUST. THE BEST MATERIAL IS 20 OZ. COPPER. ZINC AND ALUMINUM WILL WORK. SILICA-GEL, A DRYING AGENT, HAS BEEN USED EXPERIMENTALLY AS A PERMANENT TERMITE CONTROL SPRINKLED AROUND A BUILDING DURING CONSTRUCTION. IT IS EXTREMELY ABSORBENT AND WHEN IT TOUCHES AN INSECT'S SHELL IT EATS A HOLE IN IT AND THEN GOES ON TO DEHYDRATE THE CRITTER. BECAUSE INSECTS HAVE EXO-SKELETONS, THEY ARE THE ONLY ONES EFFECTED BY SILICA-GEL. IT IS TOTALLY NON-TOXIC TO FOLKS AND OTHER ANIMALS. I KNOW IT WORKS ON FLEAS IN A HOUSE. IT IS USED TO DRY FLOWERS, AND IS USED IN PACKAGING AND IN REFRIGERATION.

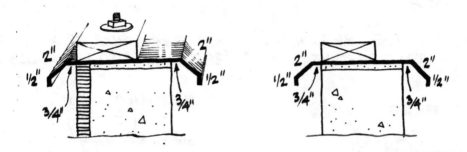

THIS TERMITE BARRIER IS THE BEST CHOICE WHERE INFESTATION IS BAD. IT OFFERS THE BEST PROTECTION BUT IS THE MOST COSTLY.

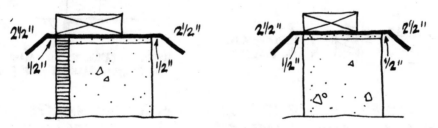

THIS TERMITE BARRIER IS GOOD FOR MOST CONDITIONS.

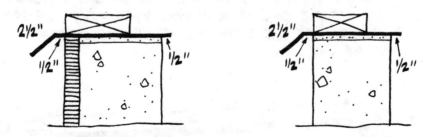

THIS TERMITE DEFLECTOR IS GOOD WHERE INFESTATION IS NOT TOO BAD. IT IS THE LEAST EXPENSIVE. USE ONLY WHEN TOP OF FOUNDATION IS EASILY VISIBLE, SUCH AS BASEMENT AREAS. DO NOT USE IN CRAWL SPACE CONDITIONS.

BENDING SHEET METAL

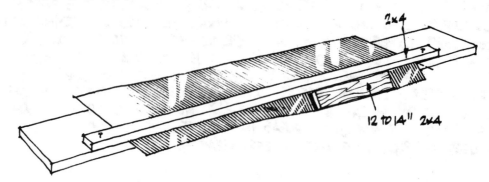

BEATING ON SHEET METAL WITH A HAMMER DOESN'T MAKE FOR A NEAT BEND. SETTING UP LIKE A SHEET METAL SHOP'S "BRAKE" HELPS DO A NEAT JOB AND IT IS QUICKER.

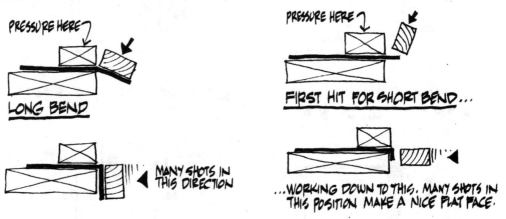

USE WHATEVER LENGTH PIECES YOU CAN HANDLE, THE SHORTER THE EASIER, IT ONLY MEANS MORE SEAMS. LOCATE THE ANCHOR BOLT HOLES, DRILL AND PUT THE FIRST PIECE IN PLACE. LAP THE NEXT PIECE (1" FOR COPPER, 1½" FOR ZINC OR ALUMINUM), LOCATE BOLTS, DRILL AND PLACE ON FOUNDATION. AT CORNERS, LAP 1" FOR COPPER AND FULL LAP FOR ZINC OR ALUMINUM. WHEN ALL PIECES ARE IN PLACE, CLEAN JOINTS AND SOLDER JUST LIKE COPPER TUBING. IF ZINC OR ALUMINUM IS USED A LOCK SEAM SHOULD BE USED.

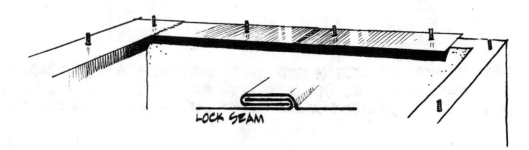

TO BEND A TERMITE BARRIER IN PLACE, USE A BLOCK OF THE REQUIRED THICKNESS TO BEND THE METAL OVER. IT IS A GOOD TWO MAN JOB, BUT NOT IMPOSSIBLE FOR ONE. ALWAYS PUT PRESSURE ON A BLOCK OF WOOD ON TOP OF THE SHEET TO KEEP IT FROM BULGING UP. MAKE BENDS BY BEATING WITH A BLOCK OF WOOD.

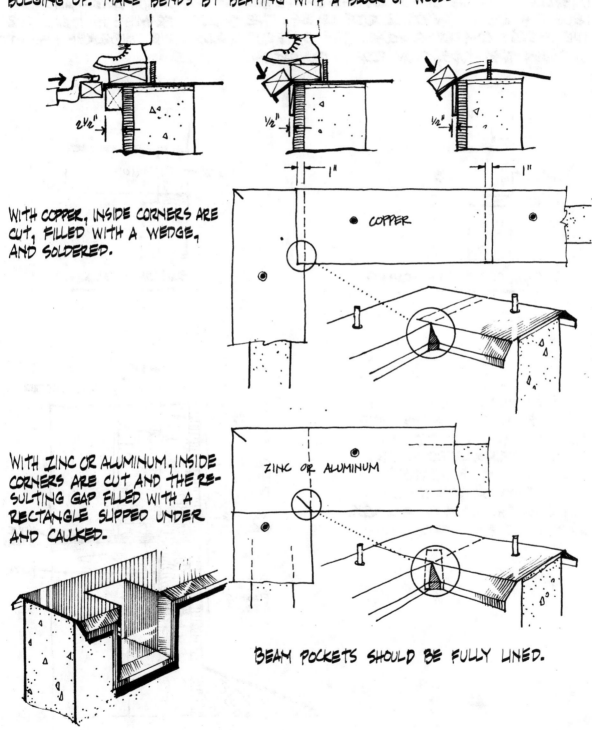

WITH COPPER, INSIDE CORNERS ARE CUT, FILLED WITH A WEDGE, AND SOLDERED.

COPPER

WITH ZINC OR ALUMINUM, INSIDE CORNERS ARE CUT AND THE RE-SULTING GAP FILLED WITH A RECTANGLE SLIPPED UNDER AND CAULKED.

ZINC OR ALUMINUM

BEAM POCKETS SHOULD BE FULLY LINED.

JOISTS AND HEADERS

THERE ARE BASICALLY TWO TYPES OF FRAMING, "BALLOON FRAMING", AND "PLATFORM OR WESTERN FRAMING". THE EXTERIOR WALL STUDS IN PLATFORM FRAMING ARE BROKEN AT EACH FLOOR BY JOISTS AND HEADERS. THE STUDS IN THE BALLOON FRAMING ARE UNBROKEN WITH THE JOISTS RESTING ON THE SILL AT FOUNDATION LEVEL, AND ON LEDGER STRIPS AT THE OTHER FLOOR LEVELS. THE BALLOON FRAMING IS IDEAL FOR THE ENERGY CONSERVING HOUSE. THE EXTERIOR WALLS HAVE UNBROKEN INSULATION, BETWEEN THE STUDS, FROM ROOF TO FOUNDATION.

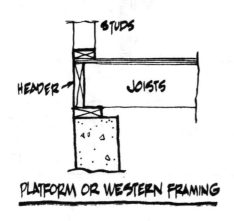

PLATFORM OR WESTERN FRAMING

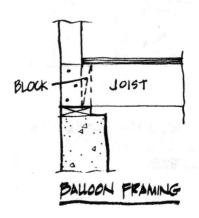

BALLOON FRAMING

AN IDEAL COMBINATION IS BALLOON FRAMING WITH THE SUPER-INSULATED WALL. IF PROPERLY BUILT, SUPERINSULATED HOUSES NEED VERY LITTLE HEATING OTHER THAN SUNLITE THROUGH WINDOWS, THE HEAT OF APPLIANCES, LIGHTS, AND BODY HEAT.

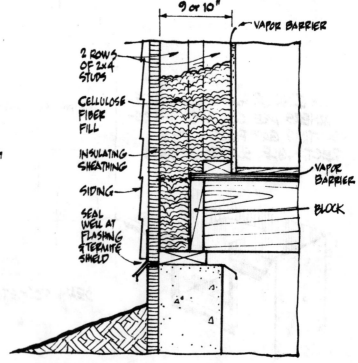

34

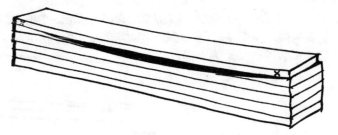

MARK AN "X" ON THE CROWN EDGE PREFERABLY ON EACH END SO YOU WON'T HAVE TO HUNT FOR IT. USE A SOFT LEAD, LUMBER CRAYON, OR FELT TIP PEN.

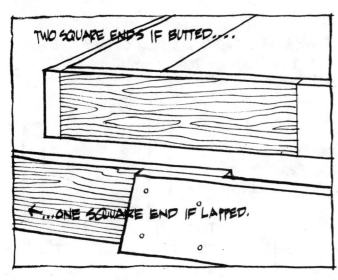

TWO SQUARE ENDS IF BUTTED.....

←...ONE SQUARE END IF LAPPED.

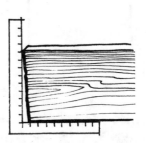

CHECK THE ENDS FOR SQUARE, BOTH ENDS IF YOU NEED THEM!

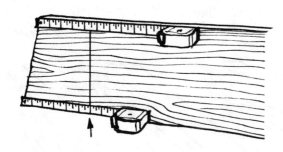

ALWAYS MEASURE FROM A SQUARED END.

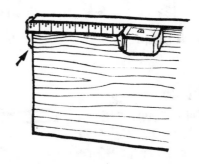

WATCH OUT FOR PIECES OF WOOD OR DIRT HANGING ON THE END, ESPECIALLY WHEN USING A TAPE MEASURE THAT HOOKS OVER THE END.

JOISTS AND HEADERS

WITH THE SILLS SQUARE AND LEVEL THE JOIST AND HEADERS SHOULD GO IN EASILY. USE STRAIGHT PIECES FOR HEADERS. IF ONE HAS A SLIGHT CROWN IT CAN BE PULLED DOWN WITH TOENAILING.

START WITH THE HEADER A LITTLE OUTSIDE THE SILL SO WHEN IT IS DRIVEN "HOME" IT WILL BE FLUSH. WHEN TOENAILING, ALWAYS HIT THE NAIL, NOT THE WOOD. LET THE NAIL DO THE PULLING.

IF THE HEADER STILL DOESN'T LINE UP WHEN THE NAIL IS DRIVEN HOME, USE ANOTHER TOENAIL. REALY POUND THOSE NAILS. IF NECESSARY USE THREE, NOT TOO CLOSE TOGETHER; THE WOOD MIGHT SPLIT.

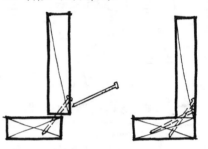

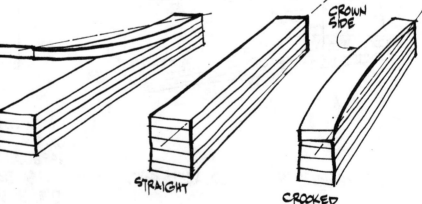

CROWN DOWN ... UP ... ... STRAIGHT ... STRAIGHT.

JOISTS SHOULD BE LAID IN CROWN UP. THE WEIGHT OF THE FLOORING WILL FLATTEN THEM OUT IN TIME. IF JOISTS ARE RANDOM CHOSEN, THERE WILL BE HIGH JOISTS NEXT TO LOW JOISTS, MAKING FOR A VERY UNEVEN FLOOR.

TO FIND THE CROWN SIGHT DOWN THE LENGTH OF THE "STICK" WHILE IT IS LYING ON THE PILE. IF YOU PICK UP ONE END IT WILL SAG MAKING A CURVED SIGHT LINE.

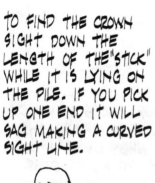

STRAIGHT

CROWN SIDE

CROOKED

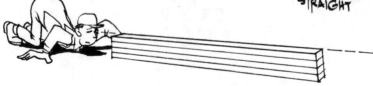

EACH END JOIST SHOULD BE STRAIGHTENED. THE BEST WAY TO SEE HOW MUCH IT IS OUT OF LINE IS TO STRETCH A STRING (MASONS LINE) FROM ONE CORNER TO THE OTHER, AT EACH END OF THE DECK, WITH ¾" SHIM BLOCKS. THE REASON FOR THE ¾" SHIM BLOCK IS TO MAKE SURE THE STRING DOES NOT TOUCH THE JOIST. SLIP ANOTHER ¾" BLOCK BETWEEN STRING AND JOIST, AT VARIOUS POINTS ALONG THE JOIST, AND YOU WILL SEE HOW MUCH OUT OR IN IT IS.

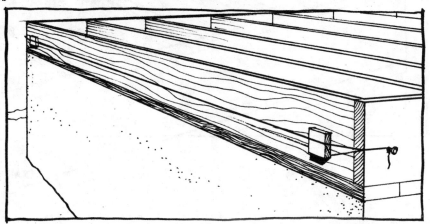

STRAIGHTEN JOIST AND HOLD IN PLACE WITH A DIAGONAL BRACE. SITTING DOWN AND PUSHING WITH THE POWERFUL LEG MUSCLES WILL MAKE ANY STUBBORN JOIST MOVE. LEAVE THE NAIL HEADS UP FOR EASY REMOVAL.

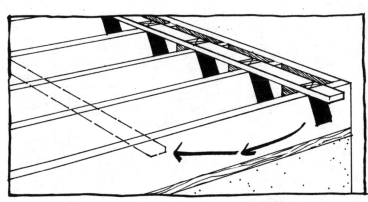

WITH THE HEADER AND END JOISTS STRAIGHT AND SECURELY BRACED, THE REST OF THE JOISTS CAN BE STRAIGHTENED. TAKE 1×3 FURRING STRIPS AND MARK OFF THE JOIST SPACINGS AT THE HEADER. THEN MOVE THE STICK 5 OR 6 FEET AWAY FROM THE HEADER AND TACK EACH JOIST IN ITS PROPER PLACE. MAKE SURE THE STRIP IS AT LEAST 50" BACK FROM THE HEADER TO ALLOW FOR ONE ROW OF PLYWOOD. STAY CLEAR OF THE BRIDGING.

IF THE SILL IS STRAIGHT THEN THE HEADER WILL BE STRAIGHT ON THE BOTTOM, AND IF THE JOISTS ARE SQUARE THEN THE HEADER WILL BE STRAIGHT ON THE TOP.

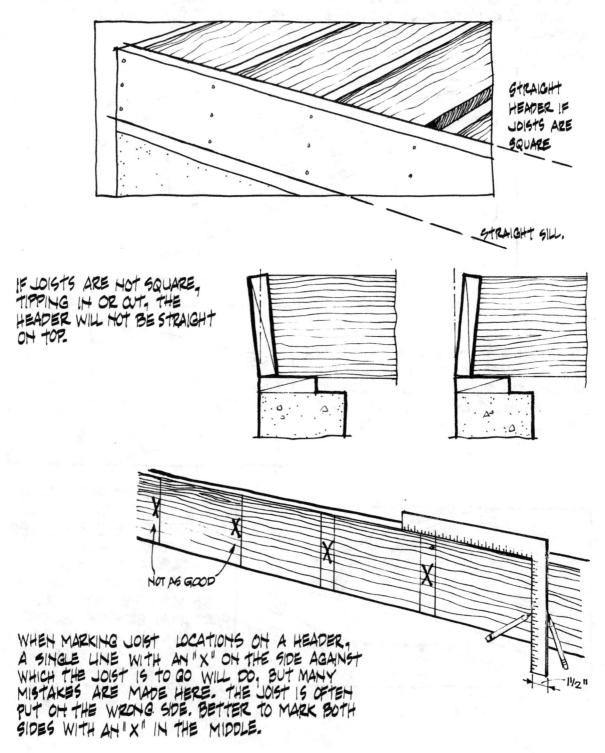

STRAIGHT HEADER IF JOISTS ARE SQUARE

STRAIGHT SILL.

IF JOISTS ARE NOT SQUARE, TIPPING IN OR OUT, THE HEADER WILL NOT BE STRAIGHT ON TOP.

NOT AS GOOD

1½"

WHEN MARKING JOIST LOCATIONS ON A HEADER, A SINGLE LINE WITH AN "X" ON THE SIDE AGAINST WHICH THE JOIST IS TO GO WILL DO, BUT MANY MISTAKES ARE MADE HERE. THE JOIST IS OFTEN PUT ON THE WRONG SIDE. BETTER TO MARK BOTH SIDES WITH AN "X" IN THE MIDDLE.

FOR JOIST INSTALLATION
BY ONE PERSON, A FOLDED-
OVER NAIL SUPPLIES SUPPORT
AT ONE END WHILE YOU NAIL
AT THE OTHER.

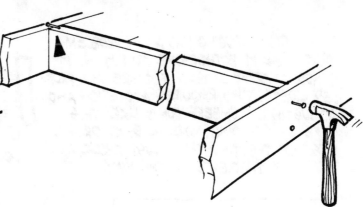

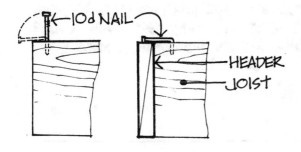

I'VE PUT IN 12 FOOT
PRESSURE TREATED
2x8'S USING THIS
SYSTEM.

SETTING JOISTS TO A
SNAPPED CHALKLINE
ON THE MUDSILL MAKES
FOR A STRAIGHT HEADER
AND EASY INSTALLATION.

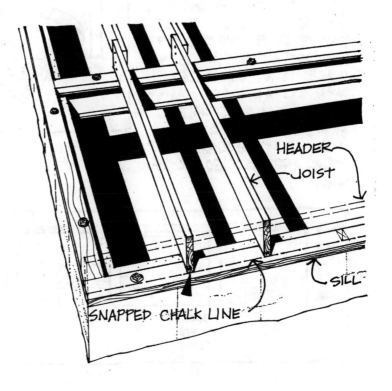

SOLID BRIDGING HAS A FEW PROBLEMS THAT MAKE IT BOTHERSOME TO PUT IN. IT TENDS TO CREEP OR GAIN IF THERE ARE PROBLEM JOISTS. THIS REQUIRES ATTENTION AND ADJUSTING. PLUMBERS DON'T LIKE THIS TYPE OF BRIDGING (MORE HOLES TO DRILL). MOST OF THE TIME THEY JUST KNOCK ANY BLOCK OUT THAT GETS IN THE WAY.

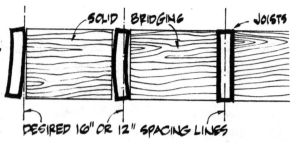

SOLID BRIDGING | JOISTS

DESIRED 16" OR 12" SPACING LINES

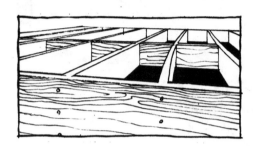

ALTHOUGH THE MATHEMATICAL SIZE OF THE BLOCKING IS 14½" (FOR 16" O.C.) OR 10½" (FOR 12" O.C.), CUT THEM A TAD SHORT, ABOUT 1/16"±, AND TEST A FEW IN PLACE. KEEP EYEBALLING DOWN THE JOISTS TO MAKE SURE THEY ARE NOT BOWING.

BRIDGING SHOULD NOT BE NAILED SOLID UNTIL THE BUILDING IS COMPLETE. THIS ALLOWS ALL THE JOISTS TO EQUALIZE AND SETTLE. WITH SOLID BRIDGING IT IS NOT EASY TO DRIVE HOME THE TOP NAIL AFTER THE DECK IS IN PLACE. IT IS TOUGH ENOUGH WITH 2X8 AT 16" O.C., IMAGINE 2X12 AT 12" O.C. YOU HAVE TO SWING THE HAMMER IN AN ARC AND AT THE RIGHT TIME (TRIAL AND ERROR) GO FOR THE NAIL.

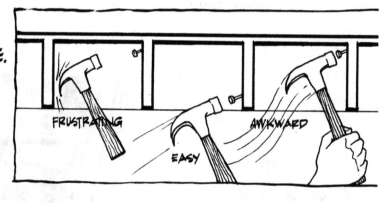

FRUSTRATING

EASY

AWKWARD

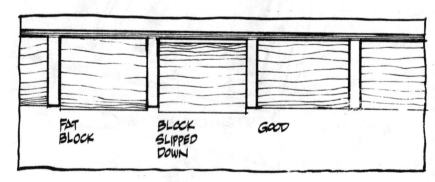

FAT BLOCK

BLOCK SLIPPED DOWN

GOOD

IF SHEETROCK IS TO GO ON THE CEILING BELOW AND NO FURRING STRIPS (STRAPPING IN NEW ENGLAND) OVER THE JOISTS, THEN THE BLOCKING MUST BE PUT IN WITH CARE. THEY MUST NOT HANG BELOW JOIST BOTTOMS. CHOP BOTTOMS OFF THOSE THAT DO WITH A HATCHET.

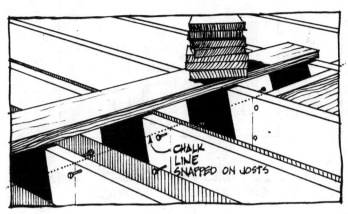

PRENAIL THE JOIST FOR SOLID BRIDG-ING. THE TOP NAIL IS NAILED WHEN YOU ARE STANDING ON A PLANK A-LONG SIDE THE CHALK LINE MARK-ING THE CENTER LINE OF THE BLOCKING. THE BLOCKS ARE STAG-GERED, SO STAGGER THE NAILS. THE BOTTOM NAILS ARE BEST NAIL-ED FROM BELOW. LOAD THE PLANK WITH BLOCKS AND NAIL THEM IN PLACE WITH THE TOP NAIL ONLY. DRIVE IT HOME HARD TO PULL THE BLOCK AGAINST THE JOIST. THE BOTTOM NAIL IS DRIVEN HOME FROM BELOW WHEN ALL BLOCKS ARE IN PLACE.

DIAGONAL WOOD BRIDGING

1X3 OR 1X4 WOOD BRIDGING CAN BE OBTAINED PRECUT AT MOST LUMBER YARDS. CUTTING IT ON THE JOB IS NOT TOO DIFFICULT ONCE YOU SET UP A JIG. THE EASIEST TOOL FOR THE JOB IS THE RADIAL SAW, BUT A SKIL-SAW AND MITER BOX SET-UP WORKS FINE. THE SIMPLEST WAY TO FIND THE SIZE IS TO DRAW IT ON A PIECE OF SCRAP JOIST MATERIAL.

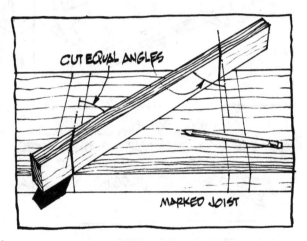

WITH THE ANGLE AND LENGTH OF THE BRIDGING KNOWN YOU CAN SET UP A MITER BOX. IF THE ANGLE IS LESS THAN 45° THE BRIDGING WILL HAVE TO BE CUT VERTICALLY BECAUSE A SKIL-SAW WILL NOT ADJUST LESS THAN 45° 1X4 BRIDGING WILL NEED A 9" SKIL-SAW TO REACH THE FULL DEPTH. .

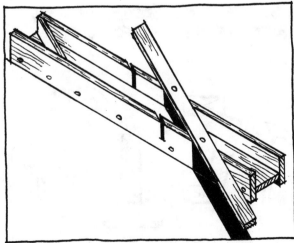

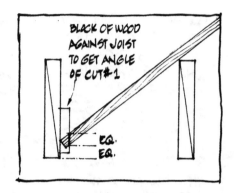

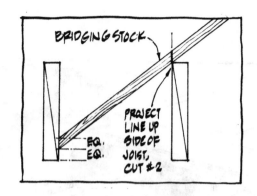

NARROW BAYS (LESS THAN 16" O.C. OR 12" O.C.) WILL HAVE TO BE CUT WITH A HAND SAW. EITHER DRAW IT ON A PIECE OF SCRAP OR MARK IT IN PLACE.

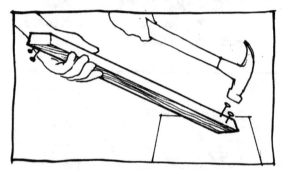

PRE NAIL THE BRIDGING WITH TWO NAILS ON TOP AND TWO NAILS ON BOTTOM.

LAY A PLANK ON THE JOISTS ALONG SIDE THE CHALK LINE, MARKING THE CENTER LINE OF THE BRIDGING. LOAD UP THE PLANK WITH THE PRE NAILED BRIDGING. NAIL THE TOP END ONLY.

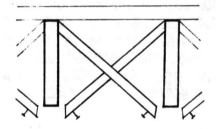

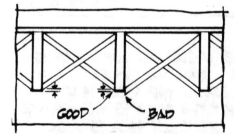

KEEP THE BOTTOMS OF THE BRIDGING UP A TAD FROM THE BOTTOM OF THE JOIST FOR BETTER BEARING.

SOMETIMES THERE IS A PROBLEM DRIVING IN A BLOCK AT AN ANGLE INTO A TIGHT PLACE. THE TOP CORNER DIGS IN AND WON'T MOVE. BEAT ON THIS CORNER WITH A HAMMER TO ROUND IT OFF. IT SHOULD AT LEAST START EASIER, BUT IF IT STILL WON'T GO, BEAT ON THE CORNER DIAGONALLY OPPOSITE.

42

PLYWOOD

EYE BALL DOWN THE JOISTS TO CHECK FOR STRAIGHT. YOU CAN PICK UP ERRORS QUICKLY THAT WAY.

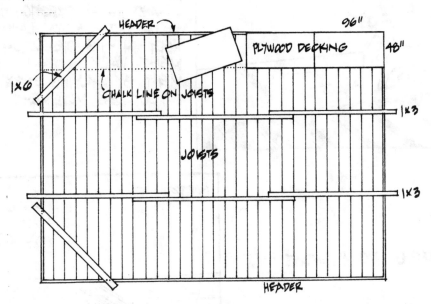

BECAUSE OF TODAY'S POOR QUALITY-CONTROL, THE PLYWOOD SHEETS SHOULD BE CHECKED FOR SIZE. THEY CAN BE AS MUCH AS 1/8" LONGER AND WIDER THAN 48" x 96". COMBINED WITH THE 6d NAIL SPACING (REQUIRED) BETWEEN SHEETS MAKES FOR QUITE A GAIN AT THE END OF FOUR SHEETS. IF YOU DON'T COMPENSATE BY GAINING A LITTLE ON THE JOIST SPACING, YOU WILL HAVE TO CUT PLYWOOD SHEETS SO THAT THEY HIT HALF WAY ON A JOIST.

SNAP A CHALKLINE ON TOP OF THE JOISTS AT 48" FROM THE HEADER AND LAY THE PLYWOOD SHEETS TO THAT LINE. THE HEADER MAY BE STRAIGHT, BUT THE CHALK LINE IS STRAIGHTER. PUT A FEW SHEETS TO WALK ON, BUT BE SURE THE ENDS ARE SUPPORTED. TACK THE FIRST ROW IN PLACE LEAVING THE NAIL HEADS UP FOR EASY REMOVAL. ONE NAIL IN EACH CORNER TO KEEP THE SHEETS FLAT.

PROCEED WITH A FEW MORE ROWS, TACKING AND SPACING. THE REASON FOR TACKING IS THAT NO MATTER HOW CAREFUL YOU ARE THE SHEETS WILL HAVE TO BE ADJUSTED TO COME HALF WAY ON A JOIST. WHEN A FEW ROWS HAVE BEEN ADJUSTED AND SET, DRIVE THE "TACKED" NAILS HOME. THIS GETS RID OF THE LITTLE TOE TRIPPERS. LAY THE REST OF THE SHEETS IN PLACE, ADJUST AND TACK. SNAP CHALK LINES ACROSS THE DECK ON JOIST CENTER LINES. NAIL IT UP.

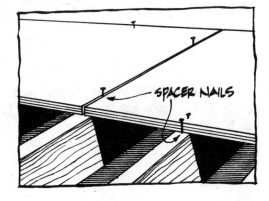

ADJUST PLYWOOD SHEETS BY JUMPING UP AND IN THE DIRECTION THE SHEET SHOULD GO. AS YOU LAND DRIVE FEET IN THE DIRECTION OF MOVEMENT. VARY INTENSITY OF MOVEMENT BY VARYING EFFORT. SMALL EFFORT LESS MOVEMENT

BIG EFFORT BIG MOVEMENT....

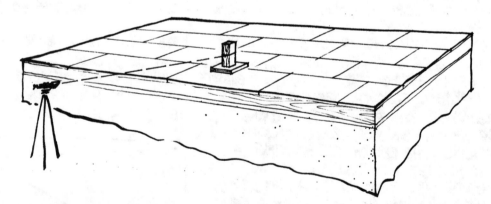

AFTER THE DECK IS ON AND NAILED IT CAN BE EASILY CHECKED FOR LEVEL WITH A BUILDERS LEVEL (TRANSIT) AND A WOOD SOLDIER. IT IS A GOOD TIME TO SET LALLY COLUMNS TOO.

T&G DECKING

T&G DECKING COMES IN 1", 2", 3" AND 4" THICKNESSES (¾", 1½", 2½", 3½"). TRY TO BUY FROM A GOOD MILL OR BE PREPARED TO STRUGGLE GETTING THE PIECES TO-GETHER. THE 3" AND 4" ARE ONLY PRODUCED BY THE BIGGER MILLS. THEY ARE ALSO DOUBLE TONGUED, SO THEY HAD BETTER FIT WELL. THEY ARE PRE-DRILLED FOR LARGE SPIKES. A HEAVY HAMMER OR A LIGHT MAUL IS A MUST.

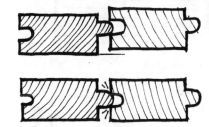

A POOR FIT IS THE RESULT OF POOR MILLING. THE TONGUE IS USUALLY SLIGHTLY LARGER THAN THE GROOVE. SOME-TIMES IT'S THE ALIGNMENT THAT IS OUT.

EVEN WITH THE VARIATIONS INVOLVED, WITH THE PROPER PERSUASION ALL WILL GO TOGETHER. A HEAVY HAMMER, PRYBARS, TOE NAILS, AND A FEW TRICKS WILL HELP. IF THE PIECE IS BOWED BADLY, TWO PRYBARS CAN BE USED ALTERNATELY TO WALK THE STICK IN. A COMBINATION OF ANY OR ALL CAN BE USED.

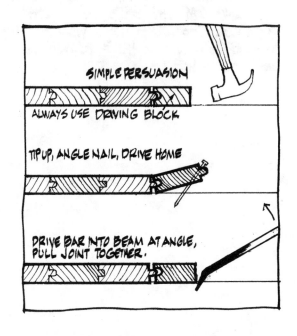

KEEP CHECKING THE ENDS FOR GAIN AND STRAIGHTNESS. BEGIN MEASURING FROM THE STARTING EDGE TO THE LAST BOARD LAID. EYEBALL DOWN THE LENGTH FOR STRAIGHT-NESS. WHEN PAST THE MIDDLE SWITCH TO MEASURING TO THE FINISHING EDGE.

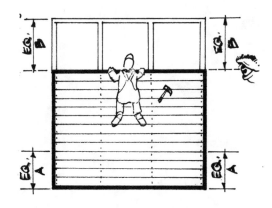

45

GAP THE WEAK END TO GAIN A LITTLE WITH EACH NEW ROW. MAKE SURE STRONG END GETS DRIVEN HOME.

IF THE MIDDLE IS BOWING OUT, DRIVE IT HARD AND GAP THE ENDS. IF A CONSTANT CHECK IS MADE, THE CORRECTIONS WILL BE MINOR. A 2" SURPRISE WITH THREE BOARDS LEFT IS NOT EASY TO MAKE UP.

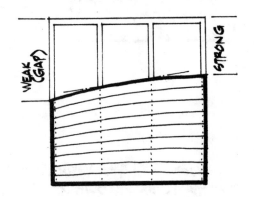

EVEN THOUGH THE SILL WAS SQUARED, ANOTHER CHECK IS IN ORDER HERE. ESTABLISH THE TRUE OUTSIDE CORNERS AND MARK THE TRUE INSIDE CORNERS USING A BLOCK OF WHATEVER THE EXTERIOR WALL PLATE IS (2×4, 2×6...). THE DECK, LIKE THE FOUNDATION, CANNOT BE CHANGED SO THE CORNERS ARE ALTERED AS WAS DONE WITH THE SILLS. SNAP A CHALK LINE CONNECTING THE CORNERS, ESTABLISHING THE INSIDE EDGE OF THE EXTERIOR WALLS. THE WALLS GO TO THIS LINE REGARDLESS OF WHAT THE DECK DOES. IF THE SILL AND JOIST WERE DONE WITH CARE, THE OUTSIDE FACE OF THE EXTERIOR WALL SHOULD LINE UP PRETTY WELL WITH THE OUTSIDE FACE OF THE DECK.

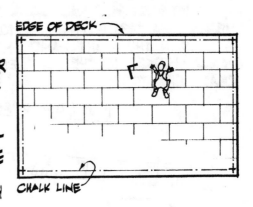

EDGE OF DECK

CHALK LINE

THE EXTERIOR WALLS ARE BUILT ON THE DECK, COVERED WITH PLYWOOD, THEN TIPPED UP INTO POSITION. IF ANYTHING OTHER THAN PLYWOOD IS USED, DIAGONAL LET-IN BRACES MUST GO IN EACH CORNER.

MOST LUMBER YARDS WILL PRE-CUT STUDS. THEY COST MORE, BUT ARE WORTH IT. A RADIAL ARM SAW ON THE JOB DOES JUST WHAT THE LUMBER YARD DOES. A SKIL-SAW WITH A JIG SET UP ON THE DECK WORKS WELL NOT ONLY FOR STUDS BUT ALSO GOOD FOR OTHER MULTIPLE CUTS. HERE IS A GOOD TIME FOR "MEASURE TWICE CUT ONCE!" I IGNORED IT ONE TIME AND CUT ALL THE EXTERIOR STUDS 12" SHORT.

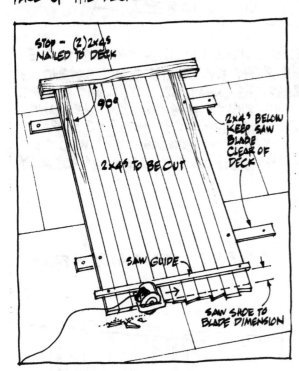

STOP - (2) 2×4'S NAILED TO DECK

90°

2×4'S TO BE CUT

2×4'S BELOW KEEP SAW BLADE CLEAR OF DECK

SAW GUIDE

SAW SHOE TO BLADE DIMENSION

I OVERCAME THAT ERROR BY BUILDING A 12" WALL USING IT AS THE WINDOW SILL HEIGHT AND THEN PUTTING THE SHORT STUD WALL ON IT.

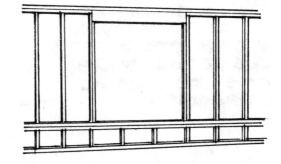

MY SHORT STUD WALL ERROR MADE ME AWARE OF THE PROBLEMS IN USING A TAPE MEASURE. A TAPE IS AN' IDEAL TOOL FOR A LEFT HANDED PERSON BECAUSE THE TAPE IS HELD IN THE RIGHT HAND AND PENCIL IN THE LEFT. THE TAPE IS THEN NATURALLY HOOKED TO THE LEFT END OF THE BOARD, EXTENDED TO THE RIGHT AND READ LEFT TO RIGHT.

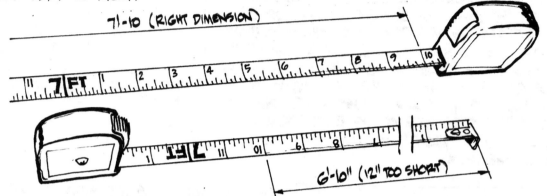

THE RIGHTY HOLDS THE TAPE IN THE LEFT HAND, PENCIL IN THE RIGHT. THE NATURAL MOTION IS TO HOOK THE TAPE ON THE RIGHT END OF THE BOARD EXTENDING THE TAPE TO THE LEFT, AND MUST THEN READ THE UNNATURAL WAY, RIGHT TO LEFT. IF THE TAPE IS HOOKED ON THE LEFT END, THE RIGHTY HAS TO TURN THE HAND OVER, EXTEND TO THE RIGHT, AND CROSS OVER WITH THE RIGHT HAND TO MARK, VERY AWKWARD. THIS BACKWARD READING CAUSES MANY ERRORS. 6½" WILL BE READ SHORT BY GOING TO THE FULL NUMBER AND THEN TO THE RIGHT ½" ACTUALLY READING 5½".

CUT TOP AND BOTTOM PLATES IN PAIRS.

THE TOP AND BOTTOM PLATES DO NOT HAVE TO BE ON A 48" MODULE TO ACCOMMODATE PLYWOOD. THEY CAN BE BUTTED ANYWHERE BECAUSE THE DOUBLE TOP PLATE AND THE LAPPED PLYWOOD WILL TIE THEM TOGETHER.

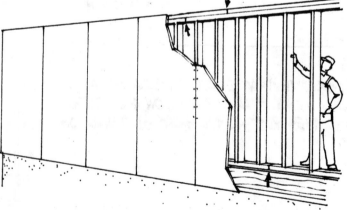

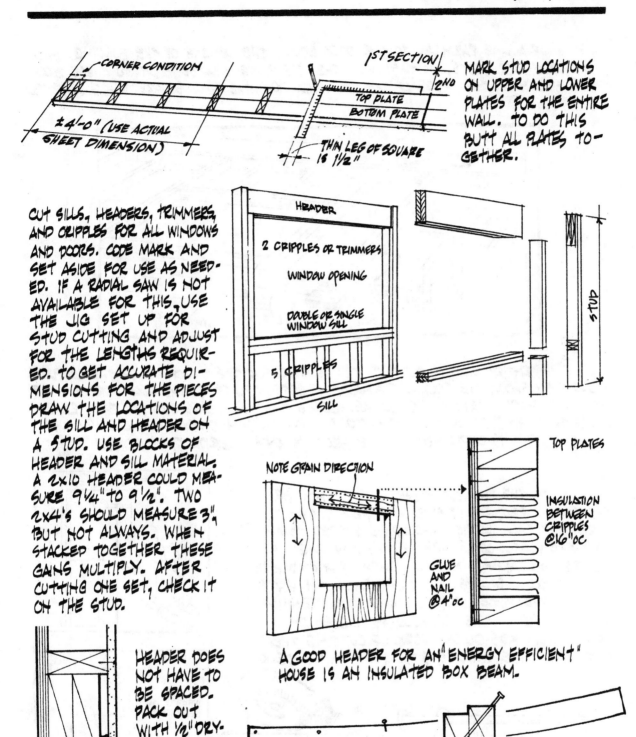

MARK STUD LOCATIONS ON UPPER AND LOWER PLATES FOR THE ENTIRE WALL. TO DO THIS BUTT ALL PLATES TOGETHER.

CORNER CONDITION

1ST SECTION
2ND
TOP PLATE
BOTTOM PLATE

± 4'-0" (USE ACTUAL SHEET DIMENSION)

THIN LEG OF SQUARE IS 1½"

CUT SILLS, HEADERS, TRIMMERS, AND CRIPPLES FOR ALL WINDOWS AND DOORS. CODE MARK AND SET ASIDE FOR USE AS NEEDED. IF A RADIAL SAW IS NOT AVAILABLE FOR THIS, USE THE JIG SET UP FOR STUD CUTTING AND ADJUST FOR THE LENGTHS REQUIRED. TO GET ACCURATE DIMENSIONS FOR THE PIECES DRAW THE LOCATIONS OF THE SILL AND HEADER ON A STUD. USE BLOCKS OF HEADER AND SILL MATERIAL. A 2x10 HEADER COULD MEASURE 9¼" TO 9½". TWO 2x4'S SHOULD MEASURE 3", BUT NOT ALWAYS. WHEN STACKED TOGETHER THESE GAINS MULTIPLY. AFTER CUTTING ONE SET, CHECK IT ON THE STUD.

HEADER
2 CRIPPLES OR TRIMMERS
WINDOW OPENING
DOUBLE OR SINGLE WINDOW SILL
5 CRIPPLES
SILL

STUD

NOTE GRAIN DIRECTION

TOP PLATES
INSULATION BETWEEN CRIPPLES @16" OC
GLUE AND NAIL @4" OC

HEADER DOES NOT HAVE TO BE SPACED. PACK OUT WITH ½" DRYWALL INSIDE.

A GOOD HEADER FOR AN "ENERGY EFFICIENT" HOUSE IS AN INSULATED BOX BEAM.

IF 2-2x4'S HAVE TO BE NAILED TOGETHER, IT IS A GOOD OPPORTUNITY TO USE UP THE CROOKED ONES BY OPPOSING THE BENDS. START AT ONE END AND WORK UP THE BOARD TO THE OTHER, TOE NAILING AS YOU GO.

SNAP A LINE PARALLEL TO THE DECK EDGE, THE LENGTH OF THE BUILDING. NAIL 2X4 BLOCKS, FOR STOPS, ON THIS LINE. THE TOP PLATE WILL BE BUILT TO THESE BLOCKS. THREE BLOCKS ON THE EDGE OF THE DECK WILL HOLD THE WALL SQUARE WHILE ASSEMBLING.

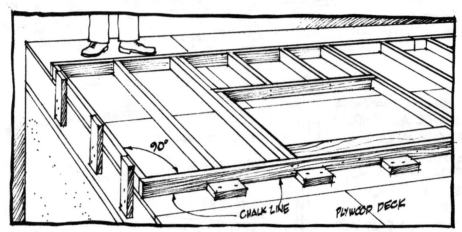

90°

CHALK LINE   PLYWOOD DECK

THE ENTIRE WALL IS BUILT ON THE DECK. CHECK THE DIAGONALS OF THE COMPLETE WALL; THEY SHOULD BE EQUAL. TOE NAIL THE BOTTOM PLATE TO THE DECK, DRIVING THE TOP PLATE HARD AGAINST THE BLOCKS. LEAVE THE NAIL HEADS OUT FOR EASY REMOVAL. IF SHEATHING OTHER THAN PLYWOOD IS USED, LET-IN DIAGONAL BRACING MUST BE USED. THIS IS EXPLAINED IN THE SECTION ON "INTERIOR PARTITIONS."

BUTT THE PLYWOOD TIGHT, ONE TO ANOTHER; THERE IS NO PROBLEM OF SWELLING FROM MOISTURE HERE, BUT EACH SECTION SHOULD BE GAPPED TO MAKE SURE THE PLATES COME TIGHT AND NO WALL GAIN OCCURS. BE CAREFUL NOT TO NAIL THE PLATES WHERE THE PLYWOOD OVERLAPS (YOU CAN'T RAISE THE ENTIRE WALL IN ONE PIECE). WINDOW AND DOOR OPENINGS ARE MARKED ON THE PLYWOOD AS IT GOES ON. A SKIL-SAW MAKES QUICK WORK OF CUTTING THEM OUT WHILE THE WALL IS STILL FLAT ON THE DECK.

MARK

SHEET 1

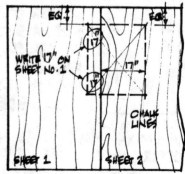

EQ.   EQ.

17   17

WRITE 17" ON SHEET NO. 1

17"

17

CHALK LINES

SHEET 1   SHEET 2

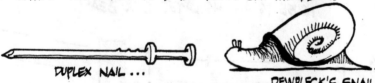

DUPLEX NAIL ...

...DEWPLECK'S SNAIL

TEMPORARY NAILING SHOULD HAVE THE NAIL HEADS OUT FOR EASY REMOVAL, BUT FOR BETTER NAIL HOLDING POWER THE DUPLEX NAIL CAN'T BE BEAT.

50

A SPRING BRACE IS A 1x6 x 16'-0"
BOARD NAILED TO THE TOP PLATE
OF A WALL TO BE STRAIGHTENED,
AND BACK TO THE DECK. THE WALL
IS PULLED IN BY PUSHING UP OR
BOWING THE MIDDLE OF THE BRACE
WITH A 50" LENGTH OF 2x4.

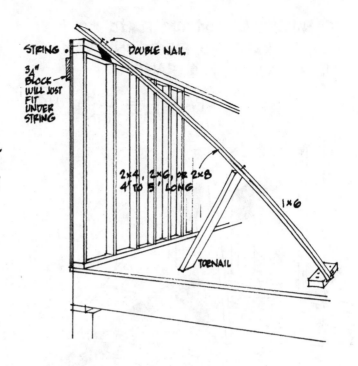

STRING

DOUBLE NAIL

3/4" BLOCK WILL JUST FIT UNDER STRING

2x4, 2x6, OR 2x8
4' TO 5' LONG

1x6

TOENAIL

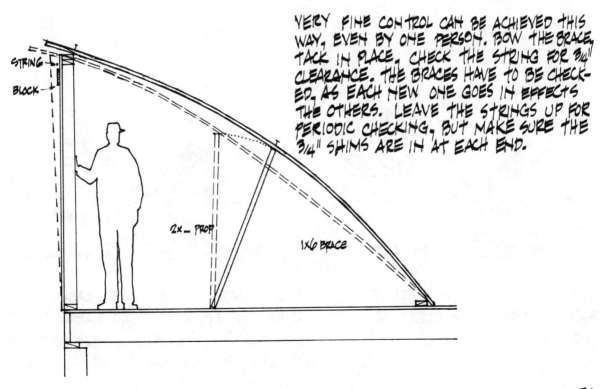

STRING

BLOCK

2x — PROP

1x6 BRACE

VERY FINE CONTROL CAN BE ACHIEVED THIS
WAY, EVEN BY ONE PERSON. BOW THE BRACE,
TACK IN PLACE, CHECK THE STRING FOR 3/4"
CLEARANCE. THE BRACES HAVE TO BE CHECK-
ED, AS EACH NEW ONE GOES IN EFFECTS
THE OTHERS. LEAVE THE STRINGS UP FOR
PERIODIC CHECKING, BUT MAKE SURE THE
3/4" SHIMS ARE IN AT EACH END.

TOENAIL THE BOTTOM PLATE TO THE
DECK TO KEEP THE PARTITION FROM
SLIDING WHEN IT'S RAISED.

CHALK LINE LOCATING
PARTITION

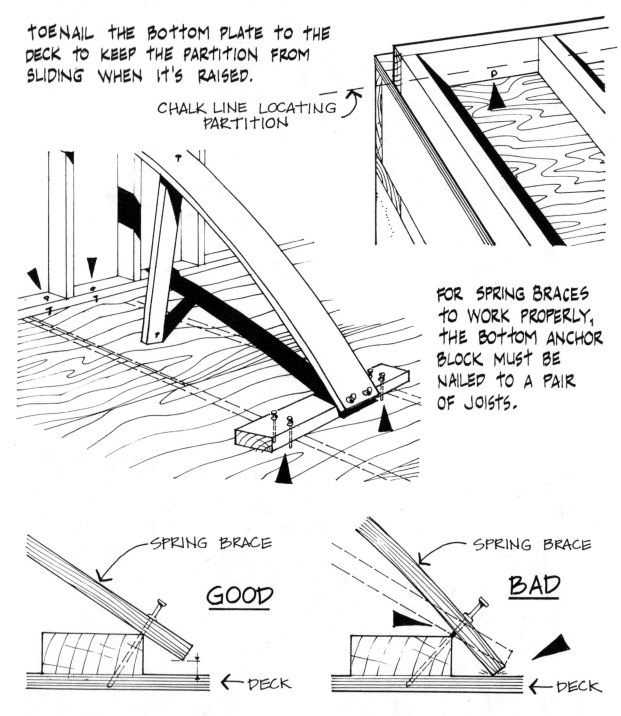

FOR SPRING BRACES
TO WORK PROPERLY,
THE BOTTOM ANCHOR
BLOCK MUST BE
NAILED TO A PAIR
OF JOISTS.

SPRING BRACE

GOOD

← DECK

THE BOTTOM OF THE SPRING
BRACE SHOULD BE CLEAR OF
THE DECK.

SPRING BRACE

BAD

← DECK

ADJUSTING THE SPRING BRACE
WHEN THE BOTTOM IS TOO CLOSE TO
THE DECK PULLS THE NAILS AND
LOOSENS THE SPRING BRACE.

BRACE THE MIDDLE SECTIONS THROUGH THE WINDOW OPENINGS. A 2X4 BLOCK NAILED THROUGH THE DECK INTO A JOIST, AND A DIAGONAL BRACE WILL DO. IF THE SECTIONS DON'T LINE UP BECAUSE OF A DIP IN THE DECK, SHIM THEM WITH WOOD SHINGLE TIPS TO BRING IN LINE.

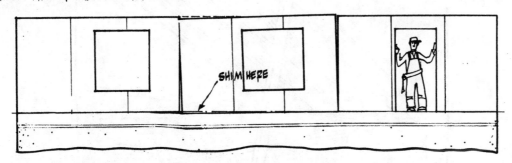

SHIM HERE

SECURE SECTIONS TOGETHER WHEN THEY ARE ADJUSTED IN PLACE AND BRACED BACK TO THE DECK, ALLOWING ROOM FOR BUILDING THE ADJACENT WALLS. BUILD THE ADJACENT WALLS THE SAME WAY. RAISE THE CORNER SECTION, DRIVE INTO POSITION AND TACK THE BOTTOM PLATE. THE CORNER IS PULLED TO-GETHER AND TACKED. CHECK FOR PLUMB EACH WAY. IT SHOULD BE ON THE MONEY IF THE PANELS ARE SQUARE AND THE DECK IS LEVEL. ONCE ALL THE WALLS ARE IN PLACE THE BOTTOM PLATE SHOULD BE NAILED SECURELY. MAKE SURE TO HIT THE JOIST AND HEADER BELOW.

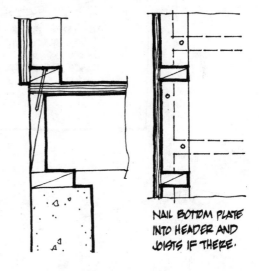

NAIL BOTTOM PLATE INTO HEADER AND JOISTS IF THERE.

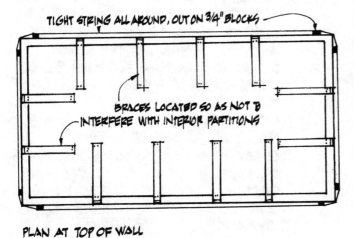

TIGHT STRING ALL AROUND, OUT ON 3/4" BLOCKS

BRACES LOCATED SO AS NOT TO INTERFERE WITH INTERIOR PARTITIONS

PLAN AT TOP OF WALL

THE EXTERIOR WALLS ARE NOW CHECK-ED FOR SQUARE WITH THE DIAGONAL MEASUREMENTS. ADJUSTMENTS ARE BEST MADE WITH A "COME ALONG" WINCH PULLING IN THE LONG DIAGONAL. FINISH ALL THE LOOSE ENDS, BOTTOM NAIL PLATES, CORNER NAIL, NAIL LOOSE AND MISSING PLYWOOD. IT IS MORE EFFICIENT TO DO NOW, LESS LIKELY TO BE FORGOTTEN. THE WALLS ARE STRAIGHTENED WITH SPRING BRACES. SHIMMED STRINGS AROUND THE PERIMETER ARE THE GUIDE LINES.

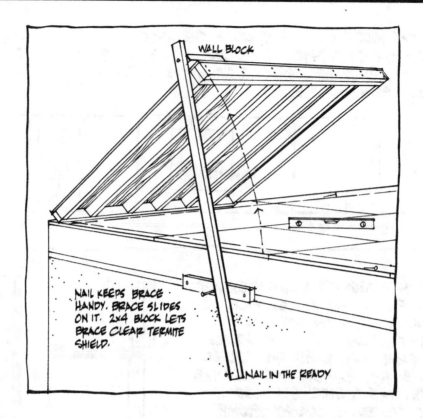

WALL BLOCK

NAIL KEEPS BRACE HANDY. BRACE SLIDES ON IT. 2×4 BLOCK LETS BRACE CLEAR TERMITE SHIELD.

NAIL IN THE READY

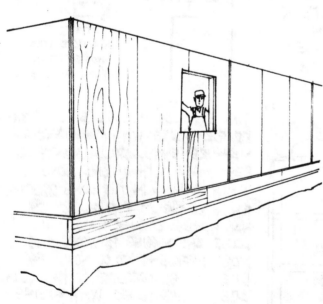

WHEN THE FIRST SECTION OF WALL IS RAISED IT WILL HAVE TO HAVE A TEMPORARY BRACE. THE BRACE MUST NOT INTERFERE WITH THE WALL THAT GOES AT RIGHT ANGLES WITH IT IN THE CORNER. A 2×4 SPACER BLOCK IS NAILED TO THE SILL BELOW AND A 2×4 BLOCK IS NAILED TO THE TOP OF THE PARTITION. TO THESE BLOCKS THE TEMPORARY BRACE IS NAILED. NAIL THE 1×3 BRACE TO THE WALL BLOCK AND PUT A NAIL AT THE READY IN THE OTHER END. A NAIL FOR THE BRACE TO SLIDE ON IS A GREAT HELP. RAISE THE WALL SECTION, DRIVE IT INTO PO-SITION WITH A SLEDGE, ROUGH PLUMB, NAIL BRACE. BEFORE RAISING WALL, CHECK THE BOTTOM PLATE FOR NAILS OR PEBBLES STUCK TO IT. TACK THE BOTTOM PLATE WITH A FEW NAILS LEAVING AN OPTION TO MOVE IT. THE OTHER SECTIONS FOLLOW.

IF EXTERIOR WALL BRACKETS ARE USED, THEY ALL GO ON AFTER THE EXTERIOR WALLS ARE SECURED STRAIGHT. THE HOLES IN THE SHEATHING FOR THESE BRACKETS ARE DRILLED IN THE FIRST STUD BAY IN EACH CORNER AND SPACED TO ACCOMMODATE THE STAGING PLANKS.

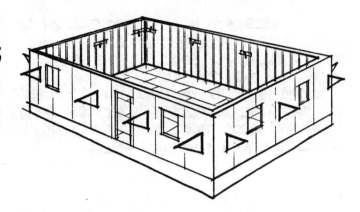

THE HOLES ARE 30" DOWN FROM THE TOP PLATE. STAGING PLANKS SHOULD BE NO LONGER THAN 14'. 12' FOOT PLANKS ARE MORE COMFORTABLE, LESS BOUNCY. DRILL HOLES LARGE ENOUGH SO THERE IS NO STRUGGLE TO GET THE BRACKET SHAFT THROUGH. NAIL A 1X3 KICKER STICK ON TOP OF THE PLANKS THAT REACHES OVER TO THE SIDING. THIS WILL KEEP THE PLANKS FROM BEING PUSHED BY A LADDER LEANED AGAINST THEM. PUT A STICK WHEREVER A LADDER WILL GO. NAIL A 1X3X18" CLEAT ACROSS BOTH PLANKS AT THE MIDPOINT.

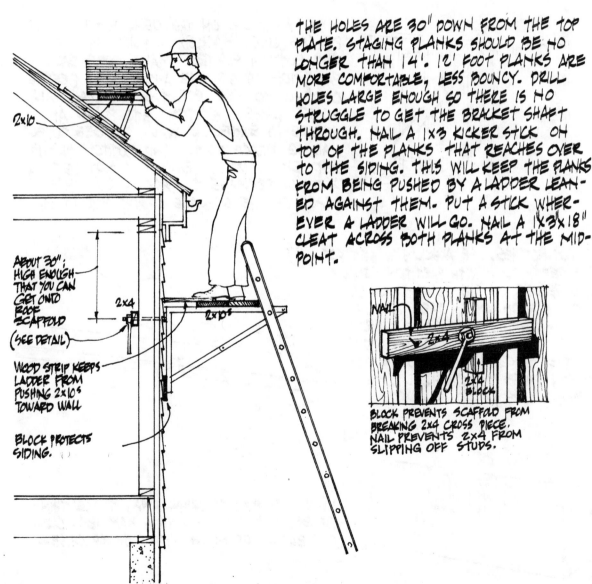

2x10

ABOUT 30". HIGH ENOUGH THAT YOU CAN GET ONTO ROOF SCAFFOLD (SEE DETAIL)

2x4

2X10s

WOOD STRIP KEEPS LADDER FROM PUSHING 2X10s TOWARD WALL

BLOCK PROTECTS SIDING.

NAIL

2x4

2x4 BLOCK

BLOCK PREVENTS SCAFFOLD FROM BREAKING 2X4 CROSS PIECE. NAIL PREVENTS 2X4 FROM SLIPPING OFF STUDS.

INTERIOR PARTITIONS

IF THERE IS TO BE NO PLYWOOD DECK ON THE SECOND FLOOR, THE EXTERIOR WALLS WILL HAVE TO BE PERMANENTLY STABILIZED. (THE SPRING BRACES ARE ONLY TEMPORARY.) THE BEST WAY TO DO THIS IS WITH A 1X3 DIAGONAL LET-IN BRACE, IN A PERPENDICULAR BUTTING PARTITION. IF THERE IS A SECOND FLOOR DECK, THESE PARTITIONS CAN BE SECURED WITH TEMPORARY BRACES.

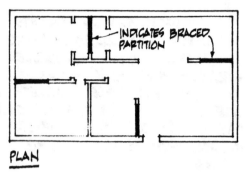

INDICATES BRACED PARTITION

PLAN

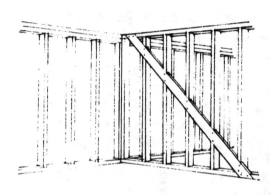

BUILD PARTITIONS ON THE DECK WITH LET-IN BRACES JUST TACKED IN PLACE AT THE TOP AND BOTTOM PLATES. MAKE SURE EACH PARTITION IS SQUARE AND SECURE BEFORE CUTTING NOTCHES FOR THE 1X3 DIAGONAL BRACE. RAISE PARTITION INTO PLACE AND SECURE TO EXTERIOR WALL BEFORE DOUBLE-NAILING BRACE TO TOP AND BOTTOM PLATE AND EACH STUD. IF BRACE IS NOT LET IN, AND NOT PERMANENT, TACK A DIAGONAL BRACE ON THE SURFACE OF THE PARTITION.

TO CUT NOTCHES, USE A SKIL-SAW SET FOR 3/4" DEPTH CUT. WITH PARTITION SQUARE AND SECURE ON THE DECK, TACK A GUIDE STRIP DIAGONALLY ON PARTITION AND CUT. MOVE THE GUIDE STRIP OVER SO THAT THE SAW WILL CUT THE WIDTH OF THE BRACE THEN CUT. CUT IN BE-TWEEN THESE TWO PARALLEL CUTS, SPACING CUTS 1/16" TO 1/8" APART.

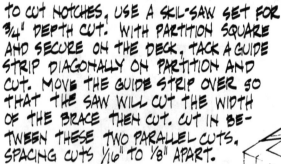

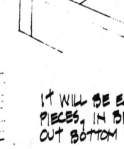

IT WILL BE EASY TO BREAK AWAY THESE THIN PIECES, IN BETWEEN, WITH A HAMMER. CLEAN OUT BOTTOM OF NOTCH WITH A SHARP CHISEL.

IF THE STANDARD FRAMING REQUIRING THREE STUDS, OR BLOCKING AND A NAILER, AT PARTITION INTERSECTIONS IS USED, A BOWED END STUD IS DESIRABLE.

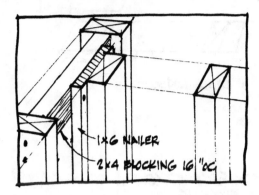

1X6 NAILER

2X4 BLOCKING 16"OC

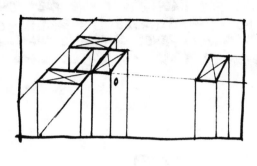

THE BOWED CONDITION FORCES THE TOP AND BOTTOM PLATES TO BE PULLED TIGHT AGAINST THE ADJOINING PARTITION. ALSO KEEP THE END STUD A TAD BACK FROM THE END OF THE TOP AND BOTTOM PLATES.

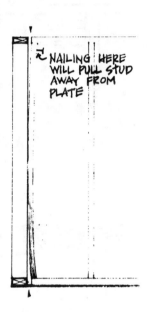

NAILING HERE WILL PULL STUD AWAY FROM PLATE

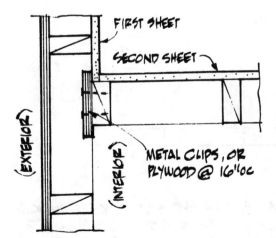

FIRST SHEET

SECOND SHEET

(EXTERIOR)

(INTERIOR)

METAL CLIPS, OR PLYWOOD @ 16"OC

IF DRY WALL CORNER CLIPS OR PLYWOOD PLATES (A FLOATING CORNER) ARE USED, THEN A STRAIGHT STUD IS A MUST. THIS FLOATING CORNER AT AN EXTERIOR WALL INTERSECTION IS IDEAL FOR THE ENERGY EFFICIENT HOUSE BECAUSE OF ITS BETTER INSULATING QUALITIES.

57

INTERIOR PARTITIONS ARE LAID OUT WITH CHALK LINES ON THE DECK. SNAP A LINE FOR EACH SIDE OF THE PARTITION. (NO GUESSING WHERE THE PARTITION GOES THAT WAY) PENCIL ANYTHING THAT SHOULD SHOW UP EASILY, AND SHOULD LAST.

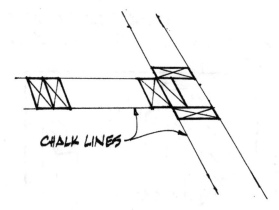

CHALK LINES

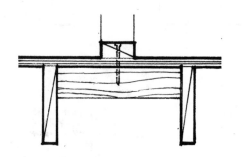

NAIL BOTTOM PLATE TO SOLID WOOD BELOW, JOISTS OR BLOCKING.

PLYWOOD ALONE WILL SAG.....

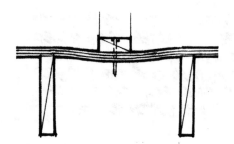

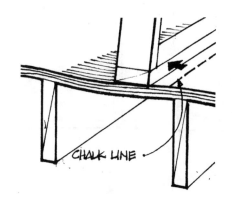

CHALK LINE

AND PARTITION MOVES OFF CHALK LINE TOO EASILY.

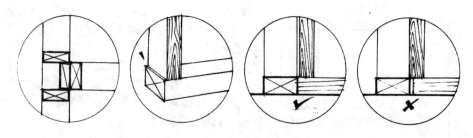

WHEN ONE PARTITION INTERSECTS ANOTHER BE SURE THE END STUD IS KEPT BACK A TAD FROM THE END OF THE TOP AND BOTTOM PLATES. IF IT IS TOO STRONG ON THE PLATES, THE PARTITION WILL NOT COME UP TIGHT. THIS IS A SIMPLE STEP THAT TAKES NO EXTRA TIME.

USE STRAIGHT STUDS FOR DOOR OPENINGS; IT WILL MAKE LIFE EASIER FOR THE "FINISH" MAN. RUN THE CRIPPLES TO THE DECK SO THAT THE BOTTOM PLATE BUTTS UP TO IT. NAIL TWO NAILS INTO THE BOTTOM PLATE, TWO AT THE TOP AND STAGGER NAIL EACH SIDE OF THE CRIPPLE IN BETWEEN. THIS WILL KEEP THE 2×4 FROM TWISTING AND IT LEAVES A MORE STABLE BASE TO SHIM THE DOOR JAMBS FROM. DOUBLE NAILING AND ALTERNATING SIDES WHEN FACE NAILING IS A GOOD WAY TO KEEP THINGS SOLID. IF JUST NAILED IN THE MIDDLE THE PIECE TENDS TO CUP AND ROCK.

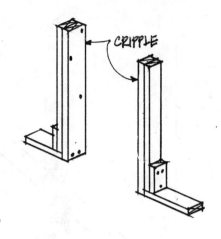

CRIPPLE

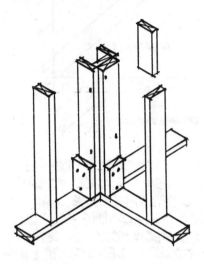

AFTER ALL THE PARTITIONS ARE IN AND THE DOOR OPENINGS CUT, NAIL 2×4 BLOCKS (6" TO 8" SCRAP) IN ALL THE CORNERS AND EACH SIDE OF THE DOOR OPENINGS. THIS WILL GIVE GOOD BACKING FOR THE BASE BOARDS.

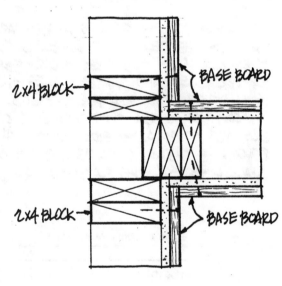

2×4 BLOCK

BASE BOARD

2×4 BLOCK

BASE BOARD

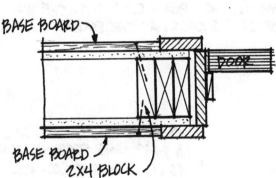

BASE BOARD

DOOR

BASE BOARD

2×4 BLOCK

IT HAS BEEN SAID "A GOOD FRAMER IS WORTH TWO FINISH MEN". IF A HOUSE IS WELL FRAMED THE FINISH WORK GOES SO EASY, BUT IF POORLY FRAMED YOU STRUGGLE AND FUDGE OVER AND OVER.

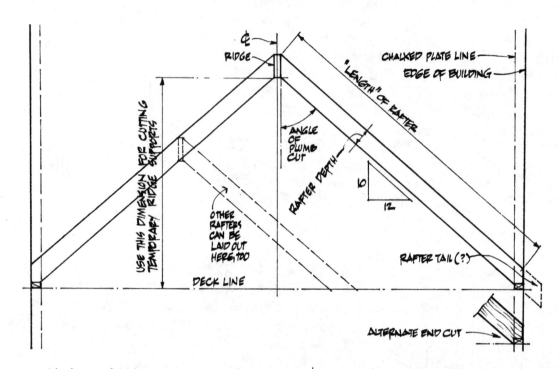

RAFTER LENGTHS AND ANGLES ARE MOST ACCURATELY OBTAINED BY DRAWING THEM ON THE DECK. IF THE SECOND FLOOR HAS NO DECK, THEN THE PLACE TO LAY THEM OUT IS ON THE FIRST FLOOR - BEFORE THE WALLS GO UP. THERE ARE MANY "RAFTER RULES" AND RAFTER TABLES THAT DO A FINE JOB. THE FRAMING SQUARE IS ANOTHER GOOD TOOL FOR THE JOB, BUT IF YOU HAVE A DECK TO WORK OFF IT IS FOOLPROOF. CHECK THAT BOTH ENDS OF THE DECK ARE EQUAL IN LENGTH. IF RAFTERS ARE TO GO ALONG SIDE JOISTS, THIS HAS ALREADY BEEN DONE WITH THE WALLS BELOW. IF THERE IS TO BE A PLATE ON TOP OF THE DECK, AT THE EDGES, SNAP A CHALK LINE FOR THE PLATE TO NAIL TO.

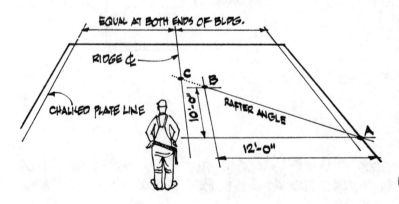

SNAP THE CENTER LINE AND A BASE LINE PERPENDICULAR TO IT. FROM THE ROOF PITCH (10 IN 12 FOR EXAMPLE) ESTABLISH THE ANGLE. STRETCH CHALK LINE FROM POINT "A" THROUGH POINT "B" CROSSING CENTER LINE AT POINT "C". THIS IS THE RAFTER ANGLE. COMPLETE THE LAYOUT.

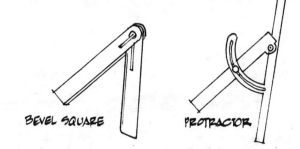

TRANSFER ANGLE, WITH BEVEL SQUARE OR PROTRACTOR, TO A PIECE OF STRAIGHT RAFTER STOCK.

BEVEL SQUARE

PROTRACTOR

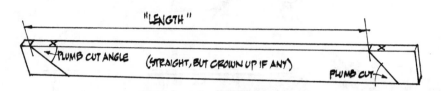

"LENGTH"

PLUMB CUT ANGLE   (STRAIGHT, BUT CROWN UP IF ANY)

PLUMB CUT

MARK AN "X" ON THE TOP TO AVOID CONFUSION EVEN IF IT IS STRAIGHT. CUT THE TOP "PLUMB CUT" ONLY ON A PAIR OF STRAIGHT PIECES OF RAFTER STOCK. CHECK HOW THEY FIT OVER THE DECK LAYOUT. WHEN THE FIT IS GOOD, LAY OUT THE REST OF THE RAFTER ON ONE PIECE ONLY. CUT AND CHECK WITH LAYOUT ON DECK. WHEN MATCH IS GOOD CUT AND MARK BOTH PIECES WITH THE WORD "PATTERN". SAVE THESE "PATTERNS" AND TWO OTHER STRAIGHT PIECES FOR THE END RAFTERS. IT WILL HELP WHEN TRIMMING WITH RAKE BOARDS.

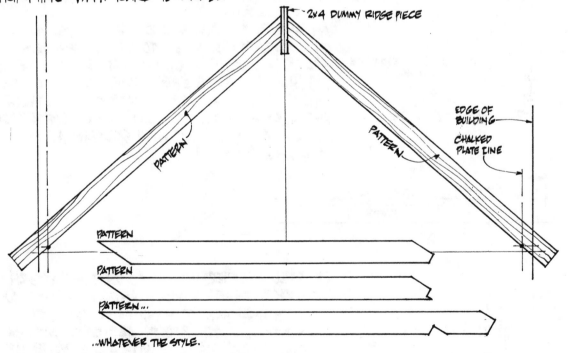

2x4 DUMMY RIDGE PIECE

PATTERN

PATTERN

EDGE OF BUILDING

CHALKED PLATE LINE

PATTERN

PATTERN

PATTERN ...

...WHATEVER THE STYLE.

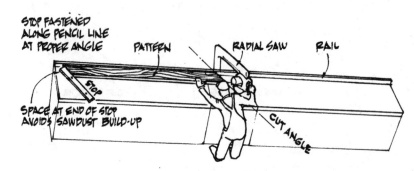

STOP FASTENED ALONG PENCIL LINE AT PROPER ANGLE

PATTERN

RADIAL SAW

RAIL

STOP

SPACE AT END OF STOP AVOIDS SAWDUST BUILD-UP

CUT ANGLE

THE RADIAL ARM SAW IS THE QUICKEST WAY TO CUT RAFTERS. USE THE "PATTERN" AS A GUIDE FOR SETTING THE ANGLE AND LENGTH.

WHEN SETTING STOPS, LEAVE A SPACE BETWEEN RAIL AND STOP. THIS KEEPS SAWDUST FROM BUILDING UP AGAINST THE EDGE WHICH WOULD CHANGE THE LENGTH OF THE CUT PIECE.

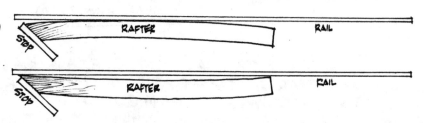

STOP   RAFTER   RAIL

STOP   RAFTER   RAIL

CUT RAFTERS WITH CROWN AWAY FROM THE RAIL. AVOID USING BAD PIECES, THEY CAUSE ALL SORTS OF PROBLEMS.

AN ANGLE "T" SQUARE AND TAPE IS ANOTHER WAY TO CUT RAFTERS. BEST TO USE A "MEASURING STICK", THE LENGTH OF THE RAFTER, FOR MEASURING (HARD TO READ THE WRONG LENGTH ON THIS STICK). THE "T" SQUARE IS MADE OF 1/4" PLYWOOD AND A 2X4. HOLD IT AGAINST THE RAFTER (OR TACK HOLD), THEN CUT WITH SKIL-SAW AGAINST THE ANGLED EDGE.

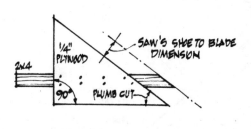

1/4" PLYWOOD

SAW'S SHOE TO BLADE DIMENSION

2X4

90°

PLUMB CUT

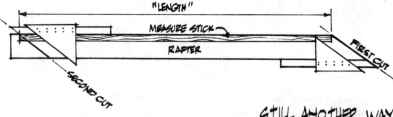

"LENGTH"

MEASURE STICK

RAFTER

SECOND CUT

FIRST CUT

STILL ANOTHER WAY IS TO USE THE PATTERN TO TRACE ON EACH PIECE AND FREE HAND CUT WITH SKIL-SAW OR HAND SAW. THE BIRD'S MOUTH AND OTHER CUTS ARE DONE WITH SKIL-SAW AFTER MARKING WITH A JIG MADE UP OF RAFTER STOCK AND PLYWOOD.
ANY COMBINATION OF THESE CAN BE USED.

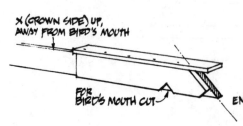

X (CROWN SIDE) UP, AWAY FROM BIRD'S MOUTH

FOR BIRD'S MOUTH CUT

END CUT

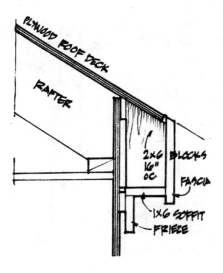

PLYWOOD ROOF DECK
RAFTER
2x6 BLOCKS 16" OC
FASCIA
1x6 SOFFIT
FRIEZE

IF A SHORT OVERHANG IS USED, BLOCKS CUT FROM THE RAFTER ENDS CAN BE USED (THE ARE THE RIGHT ANGLE TOO). THIS METHOD CREATES A NICE STRAIGHT LINE FOR THE FASCIA AND SOFFIT, AND IT DOESN'T MATTER IF THERE ARE ANY IRREGULARITIES IN THE RAFTER ENDS. IT CAN SOMETIMES MEAN USING THE NEXT SHORTER SIZE RAFTER. RUN THE BLOCKS THROUGH A TABLE SAW TO EVEN THEM UP IN LENGTH AND WIDTH. THE BLOCKS ARE NAILED ON A 1X6 OR 1X8 AT RAFTER LOCATIONS. THREE OR FOUR OF THESE SECTIONS ARE MADE UP ON THE GROUND.

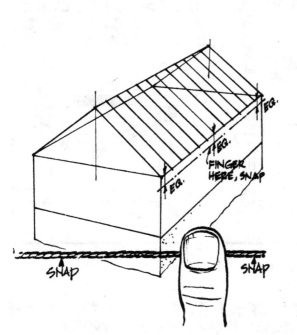

EQ.
EQ.
FINGER HERE, SNAP
EQ.
SNAP
SNAP

LOCATE THE BOTTOM OF THE BLOCKS AT EACH END AND THE MIDDLE OF THE BUILDING. STRETCH A CHALK LINE THE LENGTH OF THE BUILDING AND SNAP IT. IF THE LINE IS LONG, HOLD THE STRING AGAINST THE WALL WITH YOUR FINGER, AND SNAP EACH SIDE OF YOUR FINGERED STRING. ALWAYS FINGER A LONG LINE, PARTICULARLY ON A WINDY DAY OR IF IT IS ON A SIDEWALL OTHERWISE THE STRING WILL NOT SNAP TRUE AND A LONG LINE WILL NOT SHOW AT THE OTHER END OF THE SNAP.

USE A GAUGE BLOCK AT EACH END AND MIDDLE FOR LOCATING CHALK LINE.

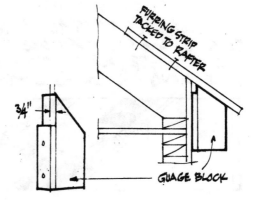

FURRING STRIP TACKED TO RAFTER
3/4"
GUAGE BLOCK

A PIECE OF 1X6 OR 1X8 FENCE BOARD WITH A "U" BRACE ON THE UPPER END, WILL SUPPORT THE RIDGE AT ONE END. ANOTHER BRACE ON THE DECK, FASTENED TO THE RAFTER SCAFFOLDING, SUPPORTS THE OTHER. AS THE RAFTERS GO UP THE SCAFFOLDING IS MOVED ALONG. OR SCAFFOLDING CAN BE BUILT THE FULL LENGTH OF THE DECK.

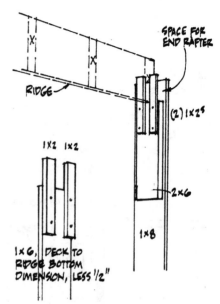

SPACE FOR END RAFTER

RIDGE

(2) 1X2'S

2X6

1X8

1X2 1X2

1X6, DECK TO RIDGE BOTTOM DIMENSION, LESS 1/2"

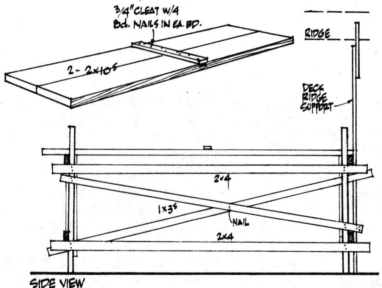

3/4" CLEAT W/4 8d. NAILS IN EA. BD.

2 - 2X10'S

RIDGE

DECK RIDGE SUPPORT

2X4

1X3'S

NAIL

2X4

SIDE VIEW

IT IS EASIER TO PUT CLEATS ON TOP OF THE PLANKS. NO PROBLEM WITH TRIPPING.

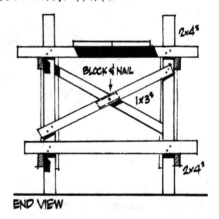

2X4'S

BLOCK & NAIL

1X3'S

2X4'S

END VIEW

CUT RIDGE PIECES TO FIT THE LENGTH OF THE BUILDING. LAY THEM ON THE DECK TO MAKE SURE THEY ARE OF THE CORRECT TOTAL LENGTH, OUTSIDE FACE OF STUD TO OUTSIDE FACE OF STUD. MARK THE RAFTER LOCATIONS BY LAYING THE RIDGE ALONG THE CEILING JOISTS, OR THE EDGE PLATE, IF USED.

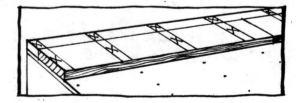

NAIL THE FIRST END RIDGE SUPPORT CENTERED ON THE FACE OF THE END WALL. THE SEAT SHOULD BE ½" LESS THAN THE HEIGHT OF THE RIDGE BOTTOM (DIMENSION ON RAFTER LAYOUT ON DECK). A HAND LEVEL FOR PLUMBING THIS SUPPORT WILL DO. THE RAFTER SCAFFOLDING WITH THE SHORT RIDGE SUPPORT NAILED AT ONE END SHOULD BE IN PLACE.

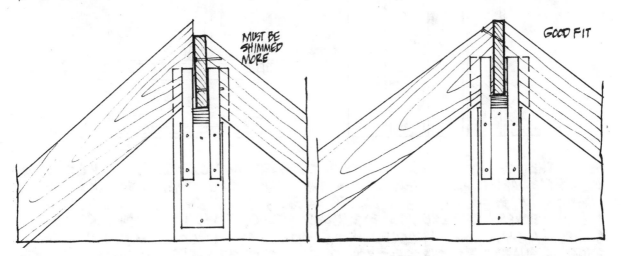

MUST BE SHIMMED MORE

GOOD FIT

WITH THE FIRST RIDGE PIECE LAID UP ON THE SUPPORTS, SHIMMED TO THE APPROXIMATE HEIGHT, ONE RAFTER IS NAILED IN PLACE. SHIM THE RIDGE TO ITS CORRECT HEIGHT WITH THE OPPOSITE RAFTER IN PLACE. MAKE SURE THE STRAIGHT END PIECES YOU SAVED, ARE USED HERE.

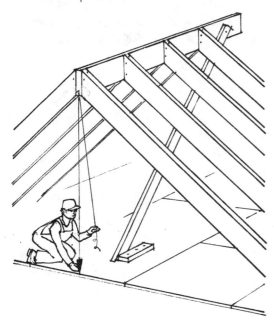

AFTER THE SECTION OF RAFTERS IS IN IS A GOOD TIME TO PLUMB THE RIDGE END, BUT PICK A CALM DAY. A MODERATE WIND IS OKAY IF THERE ARE CALM SPELLS. THE "PLUMB BOB" IS THE BEST INSTRUMENT FOR THIS JOB. LOOP THE STRING OVER A NAIL IN THE RIDGE END, CONTROL THE BOB HEIGHT WITH ONE HAND SO IT JUST CLEARS THE DECK AND STEADY IT WITH THE OTHER. WITH THE STAGING MOVED TO THE NEXT SECTION IT WILL BE EASY TO ADJUST THE FIRST SECTION TO THE PLUMB POSITION. SECURE WITH A DIAGONAL BRACE FROM RIDGE TO DECK.

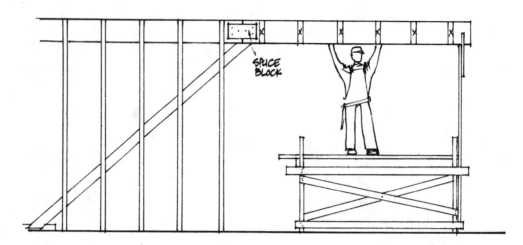

SPLICE
BLOCK

THE NEXT RIDGE PIECE IS BUTTED, TOENAILED AND CLEATED TO THE FIRST SECTION RIDGE PIECE. AFTER ALL RAFTERS ARE IN THEY ARE STRAIGHTENED JUST LIKE THE JOISTS: SHIMMED STRING ON END RAFTER FACES, STRAIGHTEN AND DIAGONAL BRACE. THE IN-BETWEEN RAFTERS ARE EYEBALLED STRAIGHT AND HELD WITH FURRING STRIPS. A CHALK LINE, FINGERED IN THE MIDDLE, AT 48" UP FROM THE RAFTER ENDS, IS SNAPPED. FOR THE FIRST ROW OF PLYWOOD TO BE NAILED TO. SINCE NO WATER CAN COLLECT HERE THE PLYWOOD SHEETS ARE LAID UP TIGHT.

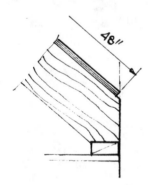

48"

TACK THE FIRST ROW TO MAKE SURE ALL IS GOING RIGHT. ONCE STARTED RIGHT THE REST OF THE SHEETS WILL GO WELL.

BACK-PRIME ALL EXTERIOR TRIM, BEFORE CUTTING TO LENGTH, TO KEEP IT FROM PICKING UP MOISTURE AND CUPPING SEVERLY. THE PAINT WILL SUCK INTO THE WOOD NICELY IF THINNED QUITE A BIT. PAINT EXPOSED ENDS AND BUTTED ENDS TO KEEP MOISTURE OUT. THE TRIM WILL LAST LONGER. DO NOT USE CLEAR PRESERVATIVE AS A PRIMER UNLESS YOU CAN WAIT A FEW WEEKS FOR DRYING. PAINT OVER PRESERVATIVE TAKES A LONG TIME TO DRY. IF RAW WOOD MUST GO UP, PUT PRESERVATIVE ON TO KEEP THE SURFACE FROM DRYING AND CRACKING. PRESERVATIVE IRRITATES THE EYES SO KEEP HANDS AWAY FROM THE FACE EVEN LONG AFTER PAINTING.

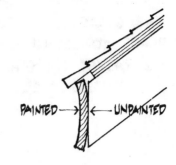

PAINTED → ← UNPAINTED

IF UNPAINTED TRIM SITS BUT A FEW HOURS IN THE SUN IT WILL SURFACE-DRY AND CUP, ULTIMATELY SPLITTING.

STACK BOARDS TOGETHER ON EDGE TO PAINT EDGES FIRST. THEN LAY THEM DOWN FOR FACES.

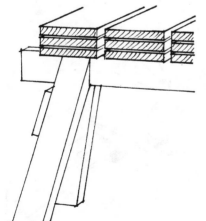

SPACE WITH THIN STRIPS OF WOOD 1/4" SQUARE OR 10d NAILS, AS MANY AS NECESSARY TO KEEP BOARDS APART.

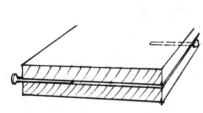

EXTERIOR TRIM

THE GARAGE IS A GOOD
PLACE TO SET UP PAINT-
ING RACKS; IT'S SPACIOUS
AND UNDER COVER.

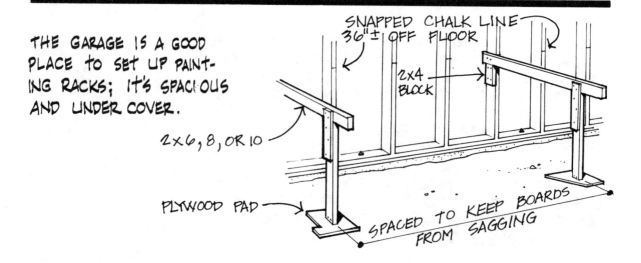

SNAPPED CHALK LINE
36"± OFF FLOOR

2×4 BLOCK

2×6, 8, OR 10

PLYWOOD PAD

SPACED TO KEEP BOARDS
FROM SAGGING

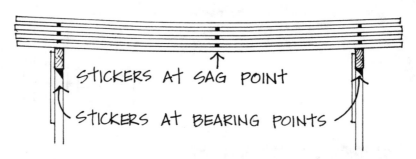

STICKERS AT SAG POINT

STICKERS AT BEARING POINTS

OIL BASE PAINT NEEDS GOOD AIR CIRCULATION. ½" SQUARE STRIPS
FOR STICKERS WORK BEST.

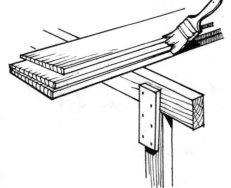

PAINTING THE EDGES OF A STACK OF
CLAPBOARDS WITH A BRUSH SPEEDS
UP THIS JOB.

PAINTING THE FACES
WITH A ROLLER (BACK
FIRST THEN TURN BOARD
OVER TO ROLL THE FACE)
REALLY SPEEDS THINGS UP.

WHEN TRIMMING THE EXTERIOR OF A BUILDING CHECK THE ENDS OF THE BOARDS FOR CRACKS. THEY WILL SOMETIMES BE VERY THIN LINES. IF THERE IS A CRACK, CUT OFF AN INCH PAST WHERE THE CRACK BEGINS.

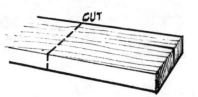

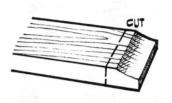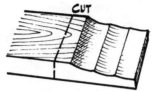

CHECK THE END FOR POOR MILLING AND CUT IT OFF.

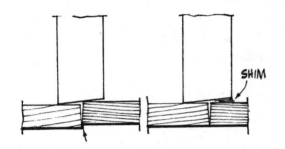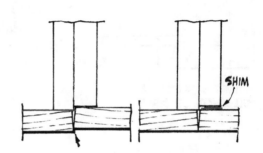

WHEN BUTTING TWO PIECES OF TRIM THE SURFACES SOMETIMES ARE NOT FLUSH. SHIM THE LOW PIECE WITH PIECES OF 15# FELT BUILDING PAPER OR WOOD SHINGLE TIPS.

TO MAKE GOOD FITTING JOINTS ON EXTERIOR TRIM RUN A HAND SAW THROUGH THE JOINT AFTER THE INITIAL CUT IS MADE AND PIECES ARE TACKED IN PLACE. IT MIGHT TAKE TWO OR THREE CUTS. FOR WOOD GUTTERS USE A VERY COARSE SAW. YOU CAN'T ALWAYS GET A SAW INTO THE JOINTS AND IN SUCH CASES IT WILL HAVE TO BE TRIMMED WITH A SHARP BLOCK PLANE.

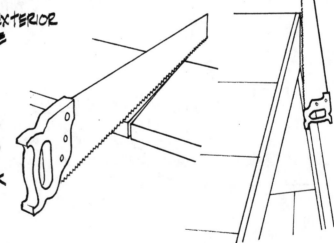

CHECK BOARDS FOR CUPPING AND PUT THE CUP IN OR HUMP OUT. THE JOINTS STAY TIGHTER THAT WAY. WOOD SHRINKS ON THE SAPWOOD OR BARKSIDE CAUSING CUPPING ON THAT SURFACE. IN CABINET WORK THE SAPWOOD IS KEPT ON THE INSIDE SURFACE OF THE WORK.

STRAIGHT BUTTED ENDS ARE CUT WITH THE VISIBLE FACE AND BOTTOM TIGHT AND THE BACK AND TOP OPEN OR RELIEVED.

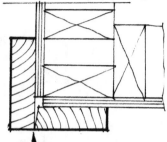

RELIEVED BACK CORNER FULL LENGTH OF BOARD

THE SAME PRINCIPLE APPLIES TO CORNER BOARDS WHERE ONE EDGE BUTTS UP TO THE OTHER FACE. USE A SLIGHT BEVEL ON A TABLE SAW OR HAND PLANE. BE SURE TO MARK THE BEVEL EDGE; IT'S SOMETIMES DIFFICULT TO TELL AT A GLANCE.

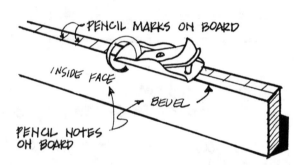

WHEN PLANING A BEVEL ON LONG BOARDS USE A 24" JOINTING PLANE.

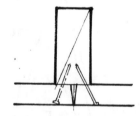

WHEN BUTTING BOARDS ON A 1½" MEMBER, USE THIN WIRE NAILS (6d COMMON OR 8d BOX NAIL), AND ANGLE THEM AWAY FROM THE JOINT.

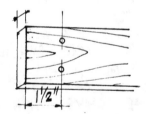

IF YOU MUST NAIL CLOSE TO THE EDGE, DULL THE NAIL POINT WITH A FEW HAMMER BLOWS. A POINTED NAIL PUSHES THE FIBERS APART CAUSING A SPLIT, BUT THE BLUNT END NAIL TEARS A HOLE AS IT GOES THROUGH.

1. PACKING STRIP FOR RAKE MEMBERS. KEEP STRIP SHORT TO ALLOW FOR EAR BOARD.
2. FASCIA.
3. GUTTER AND GUTTER RETURN WITH WATER TABLE.
4. RAKE BOARDS.
5. SOFFIT.
6. FRIEZE BOARD PACKING.
7. FRIEZE BOARD.
8. CORNER BOARDS.
9. MISCELLANEOUS TRIM.
10. RIDGE BOARDS.

1. THE RAKE BOARD IS PACKED OUT SO THAT CLAP BOARDS AND SHINGLES CAN TUCK UP BEHIND. A 1X3 FURRING STRIP IS USED IF 3/4" CORNER BOARDS ARE USED. IT IS A LITTLE TIGHT WITH SHINGLES THAT MEASURE 7/8" FROM SHEATHING TO FACE OF BUTTS, BUT IT WILL WORK (SPLIT SHAKES ARE MUCH THICKER). USE 5/4"X3 FOR 5/4" CORNER BOARDS. 3/4" WILL WORK BUT THE EAR BOARD WILL HAVE BE TRIMMED TO 3/4", BEHIND THE RAKE ONLY, TO RECEIVE IT.

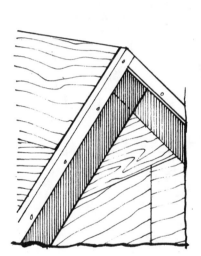

SHINGLES OR CLAPBOARDS

2. THE FASCIA IS BEVEL CUT ON TOP TO MATCH THE ROOF PITCH, AND MITERED AT ONE END FOR EAR BOARD RETURN. IF RAKE BOARD COVERS THE CORNER OF EAR BOARD AND FASCIA, MITERING IS NOT NECESSARY.

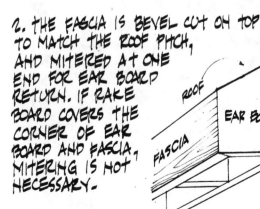

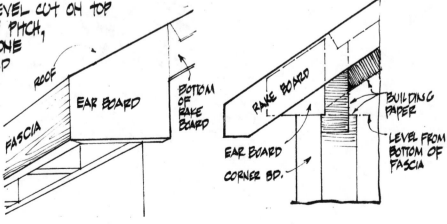

EAR BOARD SHOULD BE LEVEL ON THE BOTTOM AND SHOULD HAVE A NOTCH CUT TO RECEIVE THE TOP TONGUE OF THE CORNER BOARD. PAINT ALL CUTS.

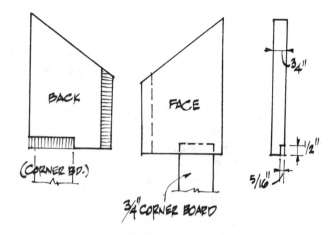

BACK

(CORNER BD.)

FACE

3/4" CORNER BOARD

3/4"

1/2"

5/16"

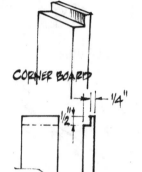

CORNER BOARD

CORNER BOARD TUCKS UNDER EAR BOARD

1/4"

1/2"

5/4" CORNER BOARD NEEDS 3/4" EAR BOARD WITH 1/4" PACKING STRIPS.

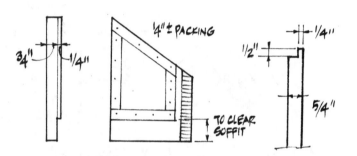

3/4"

1/4"

1/4" ± PACKING

TO CLEAR SOFFIT

1/4"

1/2"

5/4"

A STRIP OF 15# FELT IS STAPLED BEHIND RAKE AND EAR BOARD. USE A PIECE WIDE ENOUGH SO THAT PAPER ON SIDEWALLS CAN BE EASILY SLID UNDER.

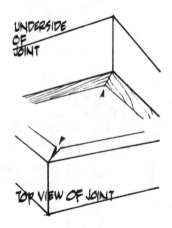

UNDERSIDE OF JOINT

TOP VIEW OF JOINT

MITERING EXTERIOR TRIM WITH A HAND SAW IS NOT DIFFI-CULT. THE OUTSIDE CORNER AND THE BOTTOM MITER LINE ARE ALL THAT IS VISIBLE. START WITH A GOOD MITER CUT ON THE BOTTOM AND FOLLOW THE OUTSIDE LINE ACROSS THE BOARD AT THE SAME TIME PUSH THE SAW OFF THE INSIDE LINE. THAT WOULD BE OPENING THE IN-SIDE OF THE MITER CUT MAKING FOR A GOOD TIGHT FIT OUTSIDE. IF THE ANGLE IS OFF ALL THAT HAS TO BE AD-JUSTED IS THE SMALL SECTION OF THE BOTTOM MITER.

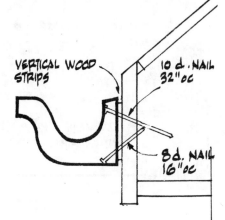

3. THE GUTTER EITHER RETURNS ON THE EAR BOARD OR BUTTS AGAINST THE RAKE. IN BOTH CASES IT SHOULD BE CUT FROM THE FASCIA WITH WOOD STRIPS TO PREVENT ROTTING.

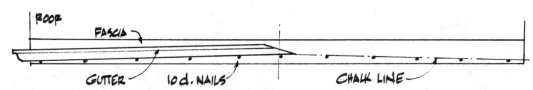

SLOPE THE GUTTER FROM MIDDLE TO ENDS ON LONG RUNS. SNAP A CHALK LINE AND THEN TACK 10d NAILS ON THE LINE TO SUPPORT THE GUTTER WHILE FITTING. SOMETIMES THE GABLE TRIM LIMITS THE SLOPE, JUST BE SURE WATER WILL NOT PUDDLE.

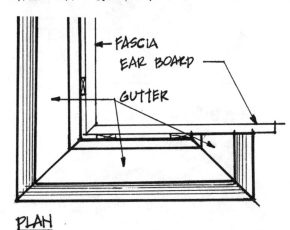

PLAN

MITER THE GUTTER AND THE GUTTER RETURNS IN A MITER BOX OR FREEHAND. MITER ONLY ONE END OF THE RETURN AND MAKE SURE IT IS PLENTY LONG. TACK IN PLACE, LEVEL, TO INSPECT FIT. RUN A COARSE SHARP SAW THROUGH THE CUT AS MANY TIMES AS IS NECESSARY FOR A GOOD FIT. NOW MITER THE OTHER END AND ITS LITTLE RETURN. THE TOP OF THE GUTTER RETURNS WILL HAVE TO BE TRIMMED FOR THE WATER TABLE TO LAY FLAT.

LEAD THE ENDS AND DRILL DOWNSPOUT HOLES WITH A HOLE DRILL IN AN ELECTRIC DRILL OR A BRACE AND BIT. A NYLON PLUMBING NIPPLE SCREWED INTO A HOLE A TAD SMALLER, WILL NOT RUST OUT LIKE A GALVANIZED NIPPLE WILL.

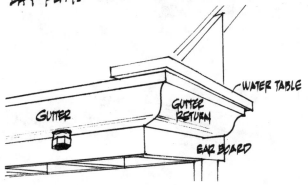

LEAD LAID IN GUTTER UP ONTO ROOF AND WATER TABLE, PREVENTS WATER FROM TURNING CORNER INTO GUTTER RETURN.

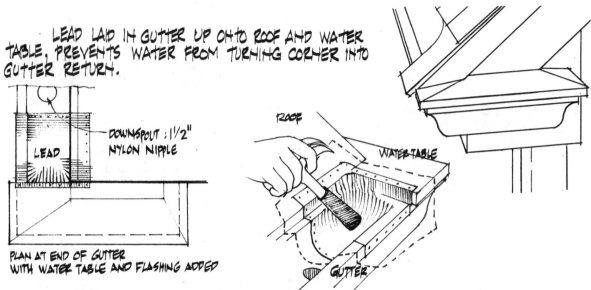

DOWNSPOUT : 1½" NYLON NIPPLE

LEAD

PLAN AT END OF GUTTER WITH WATER TABLE AND FLASHING ADDED

ROOF

WATER TABLE

GUTTER

KEEP CREASES OUT BY WORKING LEAD WITH HAMMER HANDLE. HAMMER ON WOOD BLOCK TO TUCK LEAD INTO CORNERS BEING CAREFUL NOT TO PUNCTURE LEAD. USE A LARGE PIECE OF LEAD AND TRIM TO FIT AFTER SHAPING. CAULK AND NAIL.

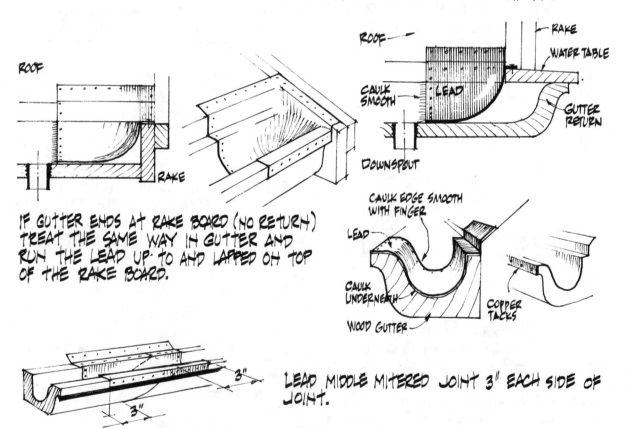

ROOF

RAKE

IF GUTTER ENDS AT RAKE BOARD (NO RETURN) TREAT THE SAME WAY IN GUTTER AND RUN THE LEAD UP TO AND LAPPED ON TOP OF THE RAKE BOARD.

ROOF

RAKE

WATER TABLE

CAULK SMOOTH

LEAD

GUTTER RETURN

DOWNSPOUT

CAULK EDGE SMOOTH WITH FINGER

LEAD

CAULK UNDERNEATH

WOOD GUTTER

COPPER TACKS

3"

3"

LEAD MIDDLE MITERED JOINT 3" EACH SIDE OF JOINT.

4. CHOOSE STRAIGHT BOARDS FOR THE RAKE IF POSSIBLE. SOME BENDING CAN BE PULLED OUT, HOWEVER. WITH DOUBLE RAKE MEMBERS, THE LAPPED PEAK IS BEST FOR TYING THE WHOLE BUSINESS TOGETHER.

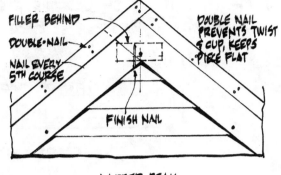

DOUBLE NAIL PREVENTS TWIST & CUP, KEEPS PIECE FLAT

FILLER BEHIND

DOUBLE-NAIL

NAIL EVERY 5TH COURSE

FINISH NAIL

LAPPED PEAK

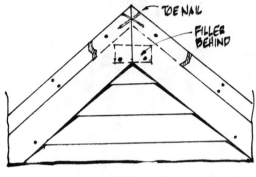

TOE NAIL

FILLER BEHIND

PLUMB CUT

PLUMB CUTS MUST BE CAREFULLY CUT AND SECURELY NAILED.

NAIL STOP BLOCKS ON ROOF AND TWO 16d NAILS INTO THE SIDE OF THE BUILDING, AT THE BOTTOM LINE OF THE RAKE BOARD. ONE PERSON CAN WORK THE RAKE BY RESTING THE BOARD ON THE NAILS. TACK IN PLACE WHILE GETTING DIMENSIONS AND FIT.

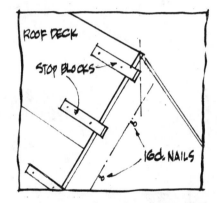

ROOF DECK

STOP BLOCKS

16d NAILS

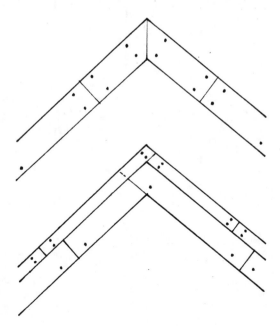

USE A SHARP BLOCK PLANE TO MAKE THE FINE FITS. IF A RAKE IS TOO LONG TO DO IN ONE PIECE, PUT THE SHORT PIECE AT THE PEAK END; IT'S NOT AS VISIBLE THAT WAY. WITH DOUBLE MEMBERS STAGGER THE JOINTS. RUN A HAND SAW THROUGH JOINTS, WHERE POSSIBLE, TO GET GOOD JOINTS. PULL RAKE UP TO BE FLUSH WITH ROOF TOP, STRONG IF ANYTHING, NOT WEAK. ASPHALT SHINGLES LOOK BAD SAGGING DOWN AND WOOD SHINGLES WILL SPLIT IF PULLED DOWN TOO FAR.

5. THE SOFFIT IS PUT ON LIKE THE FASCIA. BOTH ENDS OF EACH PIECE ARE EASED BACK (BEVELED ON THE BACK FACE).

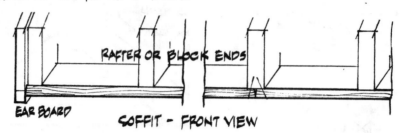

RAFTER OR BLOCK ENDS

EAR BOARD

SOFFIT - FRONT VIEW

SET ALL NAILS USED IN EXTERIOR TRIM; IT PULLS THE PIECES UP TIGHT. PUTTY NAIL HOLES BEFORE PAINTING. USE THE LARGEST NAIL SET FOR COMMON NAILS AND IF STUCK WITHOUT ONE USE A NAIL ON ITS SIDE.

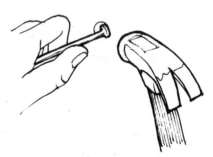

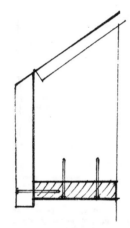

DOUBLE NAIL ON THE BOTTOM. NAIL THROUGH THE FASCIA AS NEEDED.

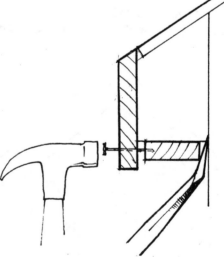

WHEREVER THE SOFFIT NEEDS PULLING IN A PRY BAR BEHIND IS SOMETIMES NEEDED TO PERSUADE THE PIECE OVER. PULL OVER, NAIL BOTTOM, THEN FRONT.

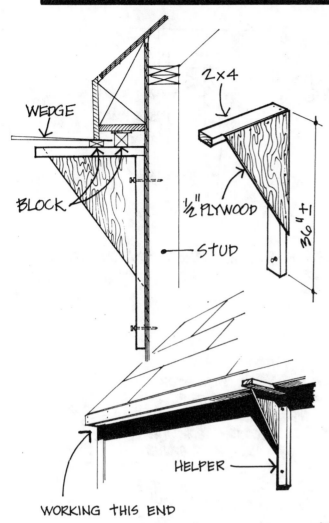

WEDGE

2x4

BLOCK

½" PLYWOOD

STUD

36"±

ONE PERSON CAN INSTALL
LONG SECTIONS OF TRIM
BY USING A SIMPLE WOOD
"L" SHAPED HELPER.

HELPER

WORKING THIS END

NAILING THE FASCIA BOARD
TO SOFFIT WHEN A CONTINUOUS
SOFFIT VENT IS USED WORKS BEST
WHEN IT IS BACKED WITH A
REMOVEABLE BLOCK SPACER
AND FORCED OVER WITH A
FLAT BAR.

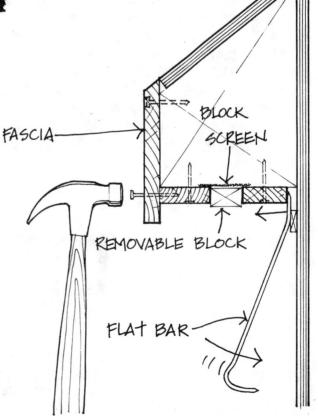

FASCIA

BLOCK

SCREEN

REMOVABLE BLOCK

FLAT BAR

6. AND 7. THE FRIEZE BOARD IS PACKED OUT WITH A 1X3 JUST LIKE THE RAKES SO THAT THE SIDING TOPS WILL FIT BEHIND. IT'S NEATER, EASIER, AND THE SHINGLES WON'T SPLIT AND FALL OUT.

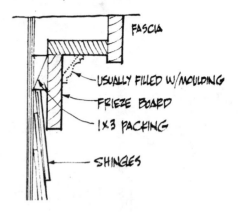

FASCIA

USUALLY FILLED W/MOULDING

FRIEZE BOARD

1X3 PACKING

SHINGLES

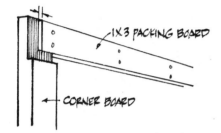

1X3 PACKING BOARD

CORNER BOARD

KEEP THE ENDS IN FROM THE CORNERS TO ALLOW FOR THE CORNER BOARD.

8. CORNER BOARDS ARE PUT ON LIKE THE RAKE BOARDS WITH STOP BLOCKS NAILED TO THE FACE OF THE BUILDING. FIRST A PIECE OF BUILD-ING PAPER (15# FELT) IS WRAPED AROUND THE CORNER EXTENDING 4" PAST EACH CORNER BOARD ON THE SIDES AND BOTTOM. FOLD THE PAPER DOWN THE MIDDLE BEFORE STAPLING UP. ALL PAPER SPLINE STRIPS SHOULD BE CUT AT THE SAME TIME. IF YOU NEED 144' OF 12" SPLINES (THE LENGTH OF A 15# FELT ROLL), CUT IT OFF THE END OF A ROLL WITH A SKIL-SAW. A 9" SAW WON'T QUITE REACH THROUGH ONE HALF OF THE ROLL. MAKE SURE ALL LAPPING PAPER HAS THE UPPER PIECE OVER THE LOWER PIECE FOR WATER RUN OFF.

15 LB. FELT SPLINE STRIPS

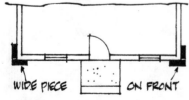

WIDE PIECE   ON FRONT

THE NARROW CORNER BOARD PIECE IS NAILED ON FIRST. THE WIDER CORNER BOARD GOES ON THE FACE OF THE BUILD-ING AS YOU LOOK AT THE FRONT DOOR FROM THE OUT-SIDE.

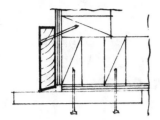

THE NARROW PIECE IS NAILED ON FIRST, KEEPING IT SNUG AGAINST THE STOPS. 8 d COMMON NAILS ARE USED FOR TRIM NAILING. WITH AN OCCASIONAL FINISH NAIL AND BRAD.

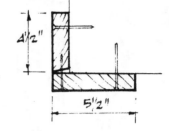

THE WIDER PIECE IS NEXT, KEEPING IT FLUSH AGAINST AND EVEN WITH THE EDGE OF THE NARROW ONE.

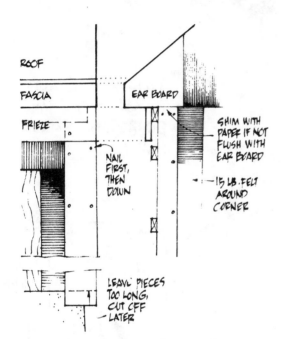

ROOF

FASCIA   EAR BOARD

FRIEZE

SHIM WITH PAPER IF NOT FLUSH WITH EAR BOARD

NAIL FIRST, THEN DOWN

15 LB. FELT AROUND CORNER

LEAVE PIECES TOO LONG, CUT OFF LATER

NAIL AT THE TOP FIRST, WORKING DOWN. THE WIDE PIECE IS ALSO NAILED AT THE TOP, WORKING DOWN AT THE CORNER EDGE ONLY. MAKE SURE THE PIECES ARE FLUSH. THEN NAIL THE EDGE THAT IS AWAY FROM THE CORNER. SET ALL NAILS TO PULL PIECES TOGETHER. SOMETIMES A 10 d COMMON IS NEEDED TO PULL THE CORNERS TOGETHER.

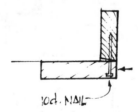

10 d. NAIL

THE WIDE BOARD USUALLY SWELLS SO A TOUCH UP WITH A SHARP BLOCK PLANE WILL BE NEEDED.

THE CORNER BOARDS ARE LEFT LONG UNTIL AFTER THE SIDING IS PUT ON. THEY ARE THEN SQUARED ACROSS WITH A COMBINATION SQUARE AND CUT WITH A HAND SAW.

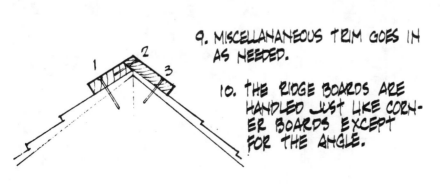

9. MISCELLANANEOUS TRIM GOES IN AS NEEDED.

10. THE RIDGE BOARDS ARE HANDLED JUST LIKE CORNER BOARDS EXCEPT FOR THE ANGLE.

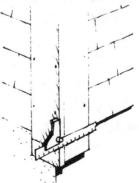

EXTERIOR TRIM

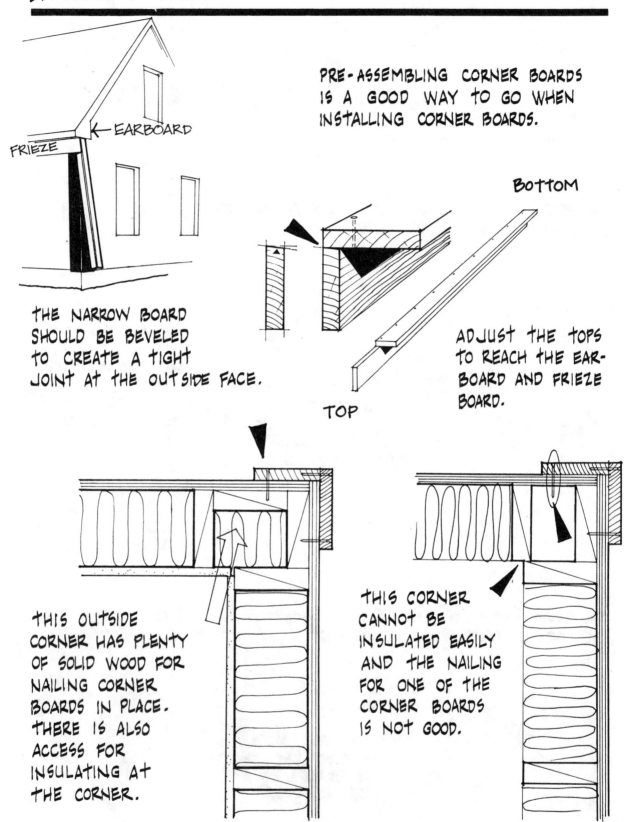

EARBOARD

FRIEZE

THE NARROW BOARD SHOULD BE BEVELED TO CREATE A TIGHT JOINT AT THE OUTSIDE FACE.

PRE-ASSEMBLING CORNER BOARDS IS A GOOD WAY TO GO WHEN INSTALLING CORNER BOARDS.

BOTTOM

TOP

ADJUST THE TOPS TO REACH THE EAR-BOARD AND FRIEZE BOARD.

THIS OUTSIDE CORNER HAS PLENTY OF SOLID WOOD FOR NAILING CORNER BOARDS IN PLACE. THERE IS ALSO ACCESS FOR INSULATING AT THE CORNER.

THIS CORNER CANNOT BE INSULATED EASILY AND THE NAILING FOR ONE OF THE CORNER BOARDS IS NOT GOOD.

WHEN BUILDING A HOUSE, GET THE ROOF ON AS SOON AS POSSIBLE, UNLESS YOU WANT THE SUMMER SUN TO DRY GREEN FRAMING LUMBER. WITH THE ROOF ON, THE BUILDING BECOMES A GOOD PLACE TO WORK AND STORE MATERIAL OUT OF THE WEATHER. WITH WINDOWS AND DOORS IT CAN BE LOCKED UP.

THIS IS A COMFORTABLE SET UP FOR ROOFING WITH EITHER WOOD OR ASPHALT SHINGLES. THE WALL BRACKET SCAFFOLDING IS THERE TO START THE ROOF FROM AND A GOOD PLACE TO KEEP SUPPLIES FOR THE ROOF WORK. LATER IT CAN BE USED FOR LOADING UP THE ROOF SCAFFOLDING WITH SHINGLES.

RED CEDAR IS THE BEST MATERIAL FOR A WOOD SHINGLE ROOF. SOME PEOPLE SUFFER SEVERE INFECTIONS FROM RED CEDAR SPLINTERS. NOTHING TO WORRY ABOUT, BUT BE AWARE.

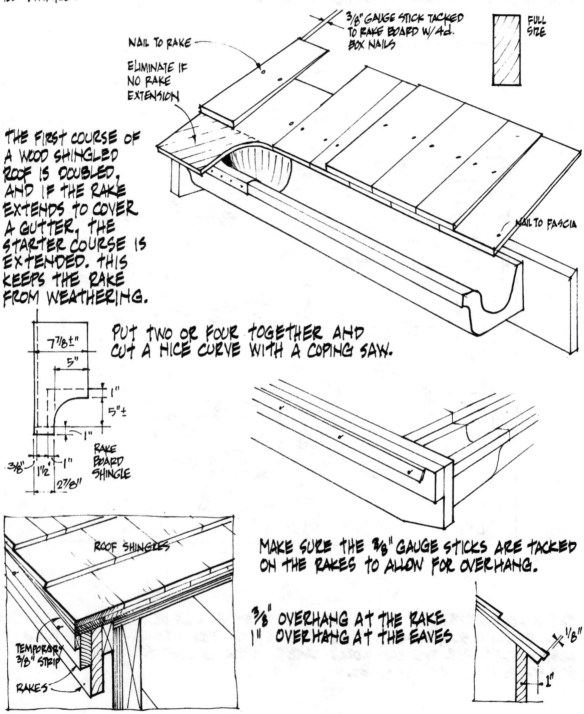

3/8" GAUGE STICK TACKED TO RAKE BOARD w/ 4d. BOX NAILS

FULL SIZE

NAIL TO RAKE

ELIMINATE IF NO RAKE EXTENSION

THE FIRST COURSE OF A WOOD SHINGLED ROOF IS DOUBLED, AND IF THE RAKE EXTENDS TO COVER A GUTTER, THE STARTER COURSE IS EXTENDED. THIS KEEPS THE RAKE FROM WEATHERING.

NAIL TO FASCIA

7 7/8" ±
5"
1"
5" ±
1"
3/8" 1 1/2" 1"
2 7/8"
RAKE BOARD
SHINGLE

PUT TWO OR FOUR TOGETHER AND CUT A NICE CURVE WITH A COPING SAW.

ROOF SHINGLES

TEMPORARY 3/8" STRIP

RAKES

MAKE SURE THE 3/8" GAUGE STICKS ARE TACKED ON THE RAKES TO ALLOW FOR OVERHANG.

3/8" OVERHANG AT THE RAKE
1" OVERHANG AT THE EAVES

1/8"
1"

NAIL THE RAKE BOARD SHINGLES AT EACH END AND ONE SHINGLE IN THE MIDDLE OF THE ROOF. ALL WILL BE POSITIONED WITH A 1" OVERHANG AT THE EAVES. STRETCH A STRING FROM END TO END AND TACK BELOW THE BOTTOM EDGE OF THE SHINGLES. THE MIDDLE SHINGLE IS TO SUPPORT THE STRING IN A LONG SPAN. KEEPING THE STRING BELOW THE SHINGLE BOTTOM LETS THE SHINGLES BE PLACED CLOSE TO, BUT NOT TOUCHING, THE STRING. IF A SHINGLE TOUCHES THE STRING THE LINE MOVES DOWN, THEN ANOTHER TOUCHES, DOWN SOME MORE. EYEBALL ALONG THE BOTTOM TO BE SURE ALL IS WELL.

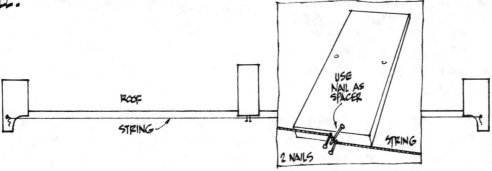

WOOD SHINGLES ARE USUALLY LAID WITH 5" TO 5 1/4" COURSES AND A ROOFING STICK WITH SHINGLE GAUGES IS MADE UP WITH THAT IN MIND. THE STICKS ARE MADE OF STRAIGHT 1X3X12', 14' OR 16' FURRING STRIPS. AS MANY AS IT TAKES TO REACH THE LENGTH OF THE ROOF.

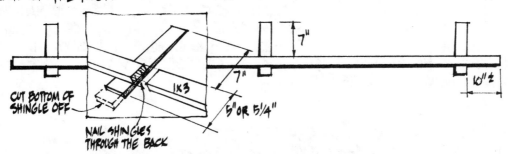

START WITH THE SECOND COURSE, USING THE ROOFING STICKS BY LINING UP THE BOTTOMS OF THE SHINGLE GAUGES WITH THE COURSE BELOW. THIS WILL GIVE AN AUTOMATIC 5" OR 5 1/4" COURSE. NAIL THE VERY TIP OF THE GAUGE TO THE ROOF WITH A WOOD SHINGLE NAIL.

THE STICK IS REMOVED BY TAPPING IT WITH A HAMMER LEAVING THE NAIL IN THE ROOF.

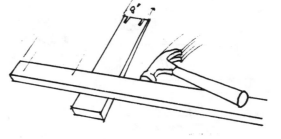

WITH THE STICKS SECURED THE FULL LENGTH OF THE ROOF JUST LAY THE SHINGLES WITH THE BUTTS AGAINST THE 1X3 AND NAIL. LAY AS MANY AS IS COMFORTABLE AND THEN NAIL, BUT DO NOT NAIL THE SHINGLE THAT IS OVER THE GAUGE UNLESS IT SPANS BY QUITE A BIT. IT WILL EITHER SPLIT OR THE GAUGE SHINGLE WILL GET NAILED. MAKE SURE THE SHINGLES DON'T SLIDE UNDER THE STICK. IT IS VERY EASY TO SKIP A COURSE, KEEP CHECKING.

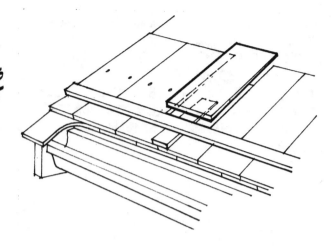

EVEN THOUGH THE SHINGLES ARE WET, SPACE THEM LOOSELY, DON'T SQUEEZE THEM IN. SOME MOISTURE WILL BE PICKED UP AND THEY WILL SWELL AND BUCKLE. IF AN OLD BATCH OF VERY DRY SHINGLES IS USED, 1/8" SPACING IS RECOMMENDED. WIDE SHINGLES SHOULD BE USED, BUT CUT, TO CONTROL THE CRACK.

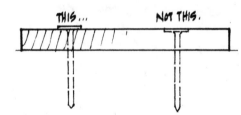

RED CEDAR SHINGLES ARE LONG LASTING AND A LITTLE HARDER THAN WHITE CEDAR. DO NOT DRIVE THE NAILS HARD, JUST ENOUGH TO SNUG THE SHINGLE TO THE ROOF. ON THE LAST STROKE, HOLD BACK A LITTLE. THE NAIL SHOULD NOT BE DRIVEN BELOW THE SURFACE OF THE SHINGLE OR THE SHINGLE WILL CRACK. BE AWARE OF CRACKED SHINGLES AS YOU USE THEM. IF SUSPECT, BEND THE SHINGLE; THE CRACK WILL SHOW. DISCARD IF BAD.

THERE IS AN UP AND A DOWN ON A SHINGLE, EYEBALL DOWN THE EDGE AND SEE. THE CURVE IS VERY SLIGHT, BUT IMPORTANT TO PUT THE HUMP UP. IF NOT, SHINGLES SKI UP.

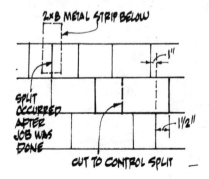

SHINGLE JOINTS SHOULD BE MINIMUM 1 1/2" OFFSET ON EACH COURSE AND 1" EVERY THIRD COURSE TO PREVENT LEAKS.

ROOF BRACKETS WITH 2X10 PLANKS FOR STAGING ARE THE MOST COMFORTABLE TO WORK OFF. LOCATE BRACKETS OVER RAFTERS FOR SECURE NAILING. AS THE STAGING IS ADVANCED UP THE ROOF BE AWARE OF THE COURSES INTO WHICH THE BRACKETS MUST GO. FIT THE SHINGLE, BUT LEAVE IT OUT. TUCK THE TIP UNDER A SHINGLE NEARBY.

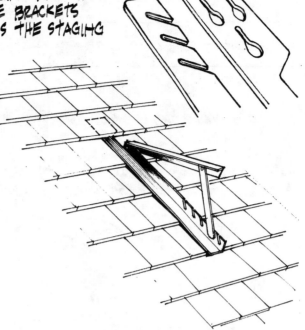

15# LB. FELT ON THE ROOF UNDER THE SHINGLES WILL FURTHER PREVENT ROOF LEAKS. ON NEW WORK, LAY ONLY AS MUCH PAPER AS IS REQUIRED FOR THE DAYS WORK. ANY EVENING MOISTURE WILL WRINKLE THE PAPER SEVERELY, MAKING IT TOUGH TO SHINGLE OVER.

CHECK DIMENSIONS FROM EAVES TO LAST COURSE, AT ENDS AND MIDDLE. ADJUST COURSE AS NECESSARY. REDUCING THE COURSE IF 16" SHINGLES ARE USED. IF 18" SHINGLES, THEN AN INCREASE WILL BE ALRIGHT. IT IS A GOOD IDEA TO SNAP A CHALK LINE EVERY ONCE IN AWHILE TO STRAIGHTEN THINGS OUT. KEEP A CHECK, AS YOU GET NEAR THE RIDGE, FOR PARALLEL AND SHINGLE COURSE DIMENSIONS. FIGURE WHERE THE BOTTOM OF THE RIDGE BOARD WILL GO, THEN MEASURE TO THE LAST SHINGLE COURSE LAID TO DECIDE ON THE ADJUSTING TO DO. A 2" LAST COURSE LOOKS BAD.

WHEN THE SHINGLING IS FINISHED, CHECK THE ROOF FOR SPLITS THAT LINE UP, AND MARK WITH A PENCIL. CUT SOME 2"x 8" METAL STRIPS TO SLIP UNDER EACH SPLIT.

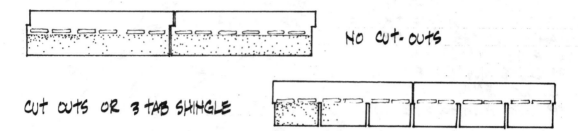

NO CUT-OUTS

CUT OUTS OR 3 TAB SHINGLE

ASPHALT SHINGLES WITHOUT CUT-OUTS ARE THE FASTER TO INSTALL AND SHOULD LAST LONGER. THE FIRST PLACE TO WEAR ON 3 TAB SHINGLES IS IN THE CUT-OUT.

WOOD SHINGLE STARTERS AT THE EAVES MAKE A NICER LOOKING JOB THAN METAL STRIPS. THE ROOF IS STARTED JUST LIKE THE WOOD SHINGLE ROOF WITH 1" WOOD SHINGLE OVERHANG. BUILDING PAPER WITH LINES IS A BIG HELP IN KEEPING THE COURSES STRAIGHT SO THE PAPER SHOULD BE PUT ON WITH CARE KEEPING EDGES RUNNING PARALLEL WITH THE EAVES.

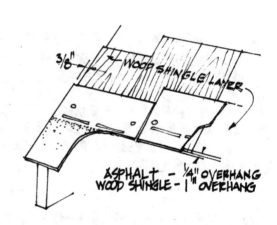

ASPHALT - 1/4" OVERHANG
WOOD SHINGLE - 1" OVERHANG

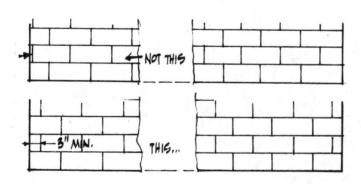

NOT THIS

3" MIN.   THIS...

MEASURE THE LENGTH OF THE ROOF SO THAT THE SHINGLE LAYOUT WILL ELIMINATE A THIN SHINGLE STRIP AT THE GABLE ENDS. WITH TAB SHINGLES, THE TABS AT EACH END LOOK BEST WHEN EQUAL.

ALL ASPHALT SHINGLES HAVE MARKS FOR ALIGNMENT. FOLLOW THE MARKS BUT KEEP AN EYE ON THE PAPER UNDER AND ADJUST TO IT PROVIDED THE PAPER HAS BEEN PUT ON PARALLEL TO THE EAVES.

ALIGNMENT SLITS

ALL THE STARTER COURSE SHINGLES ARE CUT AND INSTALLED ACCORDING TO THE INSTRUCTIONS ON THE BUNDLES. TACK THE FIRST COURSE IN PLACE, WITH A 1/4" OVERHANG ON THE WOOD SHINGLES. ACCORDING TO THE LAYOUT ARRIVED AT EARLIER. FREQUENTLY THESE LAY-OUTS TURN OUT WRONG AND ARE EASIER TO CHANGE IF THE NAILS ARE NOT DRIVEN HOME. A PIECE OF ASPHALT SHINGLE HAS TO BE CUT TO COVER THE RAKE BOARD SHINGLE. MAKE SURE THE 3/8" GUAGE STRIP IS TACKED ON THE RAKE. TOO MUCH ASPHALT SHINGLE OVERHANG TENDS TO SAG.

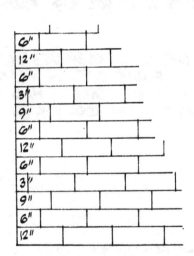

AFTER THE SHINGLE LAYOUT IS ESTABLISHED, VERTICAL LINES ARE SNAPPED FROM RIDGE TO EAVES. THESE OFF SET LINES ARE ONLY USED WITH TAB SHINGLES. THESE LINES KEEP THE CUT OUTS ALIGNED UP THE ROOF. SHINGLES WITHOUT CUT OUTS SHOULD BE RANDOM STAGGERED. THERE IS ALSO A RANDOM STAGGER FOR CUT OUT SHINGLES THAT LOOKS GOOD.

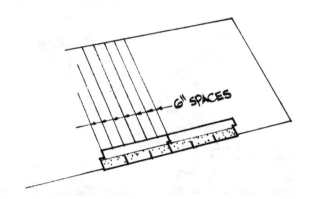

6" SPACES

ROOF BRACKETS AND 2x10 PLANKS ARE THE MOST COMFORTABLE TO WORK FROM AND GOOD FOR STACKING BUNDLES ON. AS THE RIDGE IS APPROACHED, PROVIDE FOR FULL 5" COURSES ALL THE WAY. IF THE PAPER WAS LAID PARALLEL TO THE EAVES USE IT AS A GUIDE FOR STRAIGHTNESS ALONG WITH AN OCCASIONAL EYE BALL DOWN THE COURSE. A FEW CHALK LINES FOR THE LAST FEW COURSES ARE A MUST TO BRING THEM IN LINE WITH RIDGE. DON'T BE TO FUSSY ON THE ROOF; YOU CAN'T SEE THE IMPERFECTIONS FROM THE GROUND. ONLY AT THE RIDGE IS IT NOTICABLE. BE VERY CONCERNED WITH THE THE PREVENTION OF LEAKS.

THE GABLE ENDS CAN BE EITHER CUT OFF AS THE COURSE IS WORKED ON OR CUT WHEN FINISHED ALONG A SNAPPED CHALK LINE. THE 3/8" GAUGE STRIP IS USED AS A GUIDE FOR CUTTING AS YOU GO. STARTER PIECES CAN BE CUT AND WORKED FROM ONE END. IF THEY ARE CUT AS YOU GO, LAY THE SHINGLE IN PLACE, REACH UNDER AND MARK WITH NAIL, TURN OVER OVER AND CUT WITH KNIFE OR TIN SNIPS.

DIVIDE A FULL SHINGLE INTO THIRDS TO GET THE CAP PATTERN. CUT THE ENDS AWAY FROM THE BUTT, TAPERED, SO THAT WHEN THEY FOLD OVER THE RIDGE THE

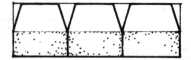

CORNERS WON'T STICK OUT. TAKE ONE OF THE CAPS AND MARK "PATTERN" ON IT. LAY IT ON THE BACK SIDE OF A FULL SHINGLE AND CUT WITH A UTILITY KNIFE. USE THE PATTERN AS A GUIDE FOR THE KNIFE. KEEP CAPS WARM FOR EASY FOLDING.

SNAP A LINE FOR ONE SIDE OF THE RIDGE AND NAIL CAPS TO IT. KEEP CAPS WARM AND DO IT ON A SUNNY DAY. THE CAPS WILL CRACK IF FOLDED OVER ON A COLD DAY. NAIL ONE SIDE OF EACH CAP AND IF IT IS WARM BEND IT OVER TO NAIL THE OTHER. IF HOT, NAIL THEM ALL ON ONE SIDE AND LET THE SUN WORK FOR YOU. BEND AND NAIL WHEN THEY ARE WARM.

IF SHINGLES MUST BE CARRIED UP A LADDER CARRY THEM ON YOUR SHOULDER.

TIP BUNDLE TO GROIN AREA. LEFT HAND IS TO THE BODY SIDE OF THE CENTER LINE OF THE BUNDLE. THE RIGHT HAND IS AWAY FROM THE CENTER LINE.

DRIVE HIPS FORWARD AND AT THE SAME TIME PULL WITH ARMS TO FINAL RESTING PLACE, THE RIGHT SHOULDER.

ANY WOOD CAN BE USED FOR SHINGLES, BUT IN NEW ENGLAND WHITE CEDAR IS THE FIRST CHOICE BECAUSE OF ITS BEAUTIFUL GREYING QUALITY. RED CEDAR AGES A NOT SO PRETTY BROWN. HAND SPLIT SHAKES ARE USUALLY RED CEDAR AND MUCH THICKER. SOME OF WHAT FOLLOWS APPLIES TO CLAP BOARDS TOO. WHITE CEDARS ARE FLAT SAWN ABOUT ³⁄₈" AT THE BUTT AND 16" LONG. WHEN BUYING SHINGLES IT IS A GOOD IDEA TO BUY BRAND NAME. THEY MIGHT COST A LITTLE MORE, BUT WORTH IT IN TIME SAVED. TRY A TRIAL BUNDLE OR TWO TO SEE HOW THEY GO UP. THEY SHOULD SIT ONE NEXT TO THE OTHER WITH AN OCCASIONAL TRIMMING REQUIRED. THEY MIGHT GO BETTER RIGHT TO LEFT OR LEFT TO RIGHT. WHEN YOU FIND A GOOD BRAND, STICK WITH THEM. THEY ARE GRADED CLEARS AND EXTRAS, THE EXTRAS BEING THE BETTER OF THE TWO.

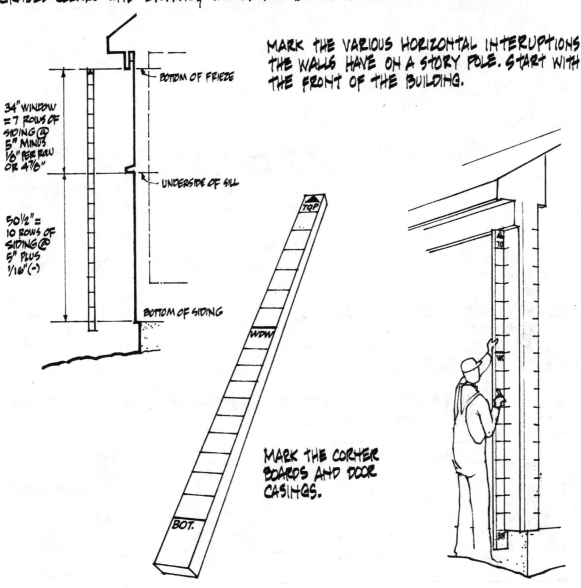

BOTTOM OF FRIEZE

34" WINDOW
= 7 ROWS OF
SIDING @
5" MINUS
⅛" PER ROW
OR 4⅞"

UNDERSIDE OF SILL

50½" =
10 ROWS OF
SIDING @
5" PLUS
1/16"(-)

BOTTOM OF SIDING

TOP

WDW

BOT.

MARK THE VARIOUS HORIZONTAL INTERUPTIONS THE WALLS HAVE ON A STORY POLE. START WITH THE FRONT OF THE BUILDING.

MARK THE CORNER BOARDS AND DOOR CASINGS.

89

IT MIGHT NEED SOME ADJUSTING AROUND THE CORNER, BUT USUALLY WINDOWS ARE SET AT THE SAME HEIGHT. THERE MIGHT BE A SHORT KITCHEN OR BATH WINDOW. SEE IF THE SAME SPACING WORKS.

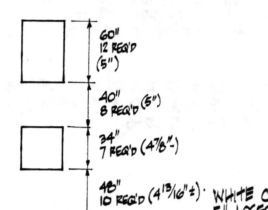

60"
12 REQ'D
(5")

40"
8 REQ'D (5")

34"
7 REQ'D (4⅞"-)

48"
10 REQ'D (4¹³⁄₁₆"±)

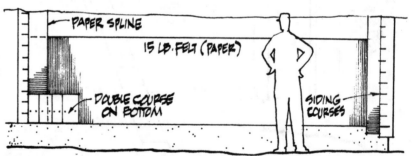

5"± COURSES

5"± COURSES

WDW.

PINE STRIP
IF 1½" OR
LESS

WHITE CEDARS ARE 16" SO STRETCHING MUCH PAST 5" LOSES TRIPLE COVERAGE, BETTER TO REDUCE. 5⅛", 5³⁄₁₆" WILL DO IF INDEED THE SHINGLES ARE 16", MANY BUNDLES HAVE QUITE A FEW "SHORTS".

START WITH BUILDING PAPER STAPLED TO THE WALL, TUCKED UNDER THE DOOR AND CORNER SPLINES. KEEP THE BOTTOM OF THE PAPER AT THE BOTTOM EDGE OF THE SHINGLES.

PAPER SPLINE

15 LB. FELT (PAPER)

DOUBLE COURSE ON BOTTOM

SIDING COURSES

THE SHINGLES CAN BE LAID TO THE BOTTOM OF THE PAPER EDGE OR ON A 1X3 STICK WITH METAL STRIPS, HUNG ON THE WALL.

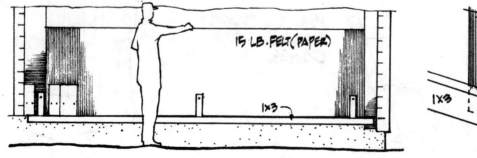

15 LB. FELT (PAPER)

1X3

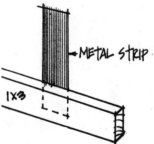

METAL STRIP

1X3

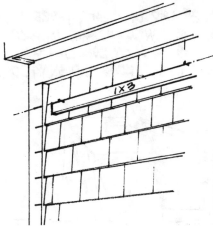

SNAP A CHALK LINE FROM CORNER TO CORNER FOR THE START OF THE NEXT COURSE. TACK ONE OR TWO 1X3 FURRING STRIPS TO THAT CHALK LINE. USE 4d BOX NAILS HIGH ON THE 1X3. THE SMALL WIRE NAIL WON'T LEAVE A BIG HOLE IN THE SHINGLE AND IF IT IS HIGH ON THE COURSE IT WILL BE LESS NOTICEABLE.

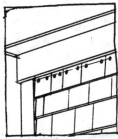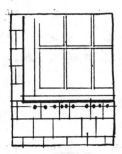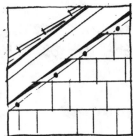

WHEN FACE NAILING SHINGLES THAT TUCK UP UNDER FRIEZE, WINDOWS, OR RAKES, USE GALVANIZED 5 d. BOX NAILS. THEY LOOK BETTER AND HOLD BETTER IF NAILS ARE IN A STRAIGHT LINE AND SET WITH A NAIL SET. THE SHADOW LINE OF THE FRIEZE, WINDOW SILL, OR RAKE CAN BE USED AS A GUIDE LINE, OTHERWISE A LIGHT PENCIL LINE, WITH A GAUGE STICK AS A GUIDE WILL DO.

SHINGLES DO NOT QUITE TOUCH

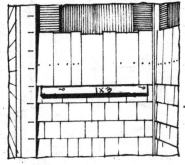

ON SHORT RUNS USE A LEVEL TO MARK THE LINE FOR A SHORT 1X3.

3/4" X 1"

TOTAL SHINGLE THICKNESS : 1"

INSIDE CORNER

FOR INSIDE CORNERS USE 3/4" X 1" TO KEEP SHINGLES BUTTS FROM CLASHING. A 1" X 1" WILL DO, BUT IT WILL BE MORE NOTICEABLE.

SIDING

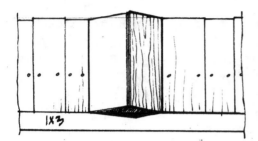

WHITE CEDARS CAN BE LAID UP TIGHT. THERE ARE NO DRY ONES, THEY WILL SHRINK. IN FACT THE TWO LAST SHINGLES IN A COURSE CAN BE ANGLED, LIKE SHUFFLING CARDS, AND SNAPPED IN.

TRIAL AND ERROR WILL SHOW HOW STRONG THE SNAP CAN BE TO MAKE THIS WORK.

snapped 'er too hard again, didn't he?

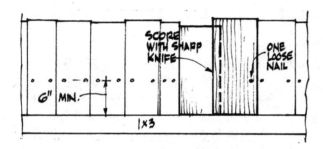

WORK BOTH ENDS TO THE MIDDLE. TWO WORKERS, A LEFTY AND A RIGHTY MAKE A GOOD COMBINATION. LAP THE LAST PAIR OF SHINGLES, SCORE WITH A KNIFE, BREAK, PUT BACK IN AND THEY SHOULD FIT. USUALLY THERE IS A SHINGLE THAT FITS PERFECTLY WITH A LITTLE SHAVE WITH A SHARP KNIFE OR HATCHET. SHINGLES SHOULD FIT AGAINST CASINGS, A BLOCK PLANE HELPS.

WHITE CEDARS HAVE IN AND OUT FACES. LOOK FOR THE CURVE AND PUT THE HUMP OUT. ITS VERY SLIGHT BUT THERE.

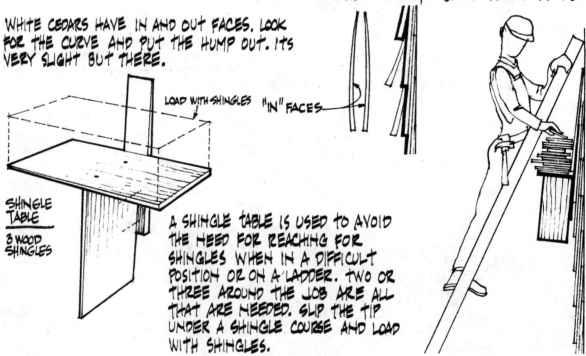

A SHINGLE TABLE IS USED TO AVOID THE NEED FOR REACHING FOR SHINGLES WHEN IN A DIFFICULT POSITION OR ON A LADDER. TWO OR THREE AROUND THE JOB ARE ALL THAT ARE NEEDED. SLIP THE TIP UNDER A SHINGLE COURSE AND LOAD WITH SHINGLES.

92

THE FIRST COURSE OF WOOD SIDEWALL SHINGLES CAN BE LAID ON A 1x3 STICK WHICH HAS A FEW SUPPORT OR HANGER SHINGLES NAILED TO IT.

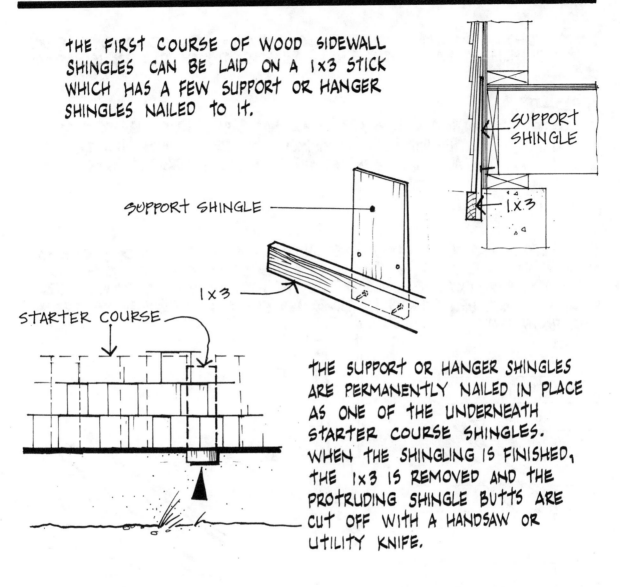

SUPPORT SHINGLE

1x3

SUPPORT SHINGLE

1x3

STARTER COURSE

THE SUPPORT OR HANGER SHINGLES ARE PERMANENTLY NAILED IN PLACE AS ONE OF THE UNDERNEATH STARTER COURSE SHINGLES. WHEN THE SHINGLING IS FINISHED, THE 1x3 IS REMOVED AND THE PROTRUDING SHINGLE BUTTS ARE CUT OFF WITH A HANDSAW OR UTILITY KNIFE.

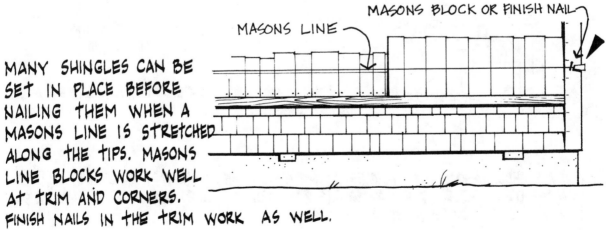

MASONS BLOCK OR FINISH NAIL

MASONS LINE

MANY SHINGLES CAN BE SET IN PLACE BEFORE NAILING THEM WHEN A MASONS LINE IS STRETCHED ALONG THE TIPS. MASONS LINE BLOCKS WORK WELL AT TRIM AND CORNERS.
FINISH NAILS IN THE TRIM WORK AS WELL.

JOINT SPACING IS NOT CRITICAL ON SIDEWALLS, BUT THEY STILL SHOULD NOT LINE UP. A 1½" MINIMUM AT EACH COURSE SHOULD BE HELD. THE THIRD COURSE IS NOT TOO CRITICAL.

SILL

MIN. ½" TO MAX. ⅝"

WINDOW SILLS ARE USUALL NOT RABBETED ENOUGH TO RECEIVE THE SHINGLES. ½" MINIMUM; ⅝" BETTER. WIDEN THE RABBET ON ALL THE WINDOWS AT THE SAME TIME BEFORE INSTALLING THEM.

USE NARROW SHINGLES AGAINST CORNERS AND CASINGS. WIDE SHINGLES SHRINK TOO MUCH LEAVING GAPS AT THESE POINTS. BE AWARE OF BAD SHINGLES, SPLITS, AND BAD SPOTS, AND VERY HARD ONES. DISCARD THE HARD SHINGLES, THEY WILL CURL BADLY AND SPLIT. SPLITS IN WHITE CEDARS ARE GENERALLY THROW-AWAYS, BREAK AND THROW AWAY.

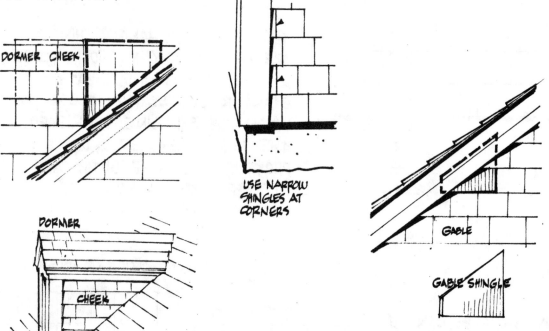

DORMER CHEEK

USE NARROW SHINGLES AT CORNERS

DORMER

CHEEK

CHEEK SHINGLE

GABLE

GABLE SHINGLE

SAVE WIDE SHINGLES ("BEDSHEETS") FOR GABLES AND DORMER CHEEKS.

CUT ALL CHEEK AND GABLE SHINGLES AT THE SAME TIME, IT'S EASY ENOUGH TO FIGURE HOW MANY. A TABLE OR RADIAL SAW IS BEST FOR THIS JOB, BUT A HAND SAW OR SKIL-SAW WILL CUT THROUGH A STACK OF THEM.

TO PUT IN OR REPLACE A SHINGLE IN A COMPLETED SHINGLE WALL USE A WOOD WEDGE, NAIL SET, AND BOX NAILS. DRIVE THE SHINGLE UP TO ½" FROM ALIGNING WITH BUTTS. USE A BLOCK, HELD AT THE BUTT, TO HAMMER AGAINST. PRY UP THE SHINGLE ABOVE AND HOLD WITH WEDGES. DRIVE TWO 4d NAILS AT AN ANGLE. SET NAILS AND DRIVE SHINGLE TO PROPER ALIGNMENT. THE UPPER SHINGLE WON'T BEND EASILY. SHINGLING UP THE GABLE END USING METAL WALL BRACKETS WILL LEAVE A CONDITION LIKE THIS AT EACH BRACKET.

TO CUT A QUANTITY OF SPECIFIC LENGTH SHINGLES (FOR UNDER WINDOWS AND FRIEZE BOARDS), CUT THEM, WHILE STILL IN THE BUNDLE, WITH A SKIL-SAW. DROP THE BUNDLE ON ITS END TO EVEN THE BUTTS. MEASURE SIZE ON TOP OF BUNDLE, AND COUNT THE LAYERS COMBINED WITH WIDTH OF BUNDLE FOR QUANTITY. 8 LAYERS AT 20"=160" OR ABOUT 13!

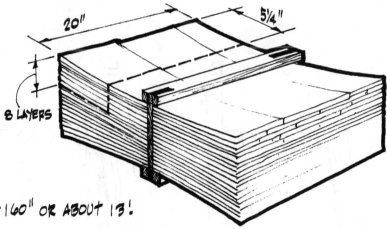

20"   5¼"

8 LAYERS

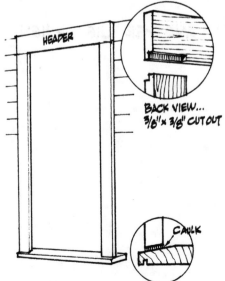

HEADER

BACK VIEW...
3/8" x 3/8" CUT OUT

CAULK

TO KEEP CASINGS WATER TIGHT, THE SIDE CASINGS SHOULD BE RABBETED INTO THE HEADER. THE SILLS SHOULD BE CAULKED.

A CLAPBOARD JIG IS A MUST FOR GETTING GOOD FITS AT CORNERS, WINDOW TRIM, AND DOOR TRIM. THE JIG STRADDLES THE CLAPBOARD AND IS HELD HARD AGAINST THE TRIM WHILE MARKING A GUT LINE WITH A UTILITY KNIFE OR SHARP PENCIL.

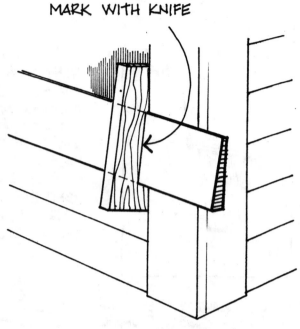

MARK WITH KNIFE

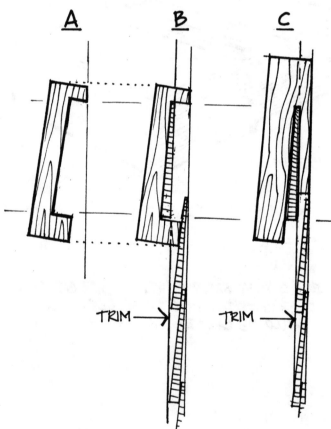

A   B   C

TRIM →   TRIM →

Ⓐ - THE JIG

Ⓑ - Ⓐ THE JIG STRADDLING THE CLAPBOARD AT CORNER OR TRIM

Ⓒ - ANOTHER JIG DESIGN

SECURE THE BUILDING AT THE END OF THE DAY. TAKE LADDERS DOWN TO PREVENT THE WINDS FROM TOPPLING THEM AND TO KEEP KIDS FROM CLIMBING UP. BATTEN DOWN THE PAPER THAT'S ON THE BUILDING. THAT INCLUDES ALL THE SPLINES AT CORNERS, WINDOWS AND DOORS AS WELL AS ROOF AND SIDEWALLS. A STRONG WIND WILL RIP THEM OFF IN A HURRY. WOOD SHINGLE TIPS, SCRAPS, FURRING STRIPS OR ANY SCRAP STICKS TACKED TO THE BUILDING WILL DO THE JOB. TUCK LOOSE SHINGLES, ASPHALT AND WOOD, UNDER UNOPENED BUNDLES. REBIND PARTLY USED WOOD SHINGLE BUNDLES. PICK UP PAPER WRAPPINGS AND WOOD SHINGLE SCRAPS; THEY LOVE TO FLY WITH THE WIND TO THE NEIGHBOR'S YARDS. THERE IS A BONUS TO THIS CLEAN UP; MANY A LOST TOOL IS FOUND THAT WAY.

LOOK OVER WHAT WAS DONE AND PLAN THE NEXT DAY'S WORK. TRY TO COMPLETE EACH JOB STAGE, NAILING OFF THE DECK, BRIDGING, WALL STRAIGHTENING, AS YOU GO. IT IS TOO EASY TO FORGET WHERE YOU LEFT OFF. IT BREAKS CONTINUITY TO HAVE TO COME BACK TO DO THIS, AND IT IS LESS EFFICIENT. MOST OF THE TIME, POSTPONED WORK DOES NOT GET DONE. CHECK FOR INCOMPLETE JOB STAGES AND PLAN TO FINISH THEM THE NEXT DAY. CHECK THE MATERIAL ON HAND, AND THE MATERIAL REQUIRED FOR THE NEXT DAY'S WORK. DOUBLE CHECK FOR TOOLS LEFT ON THE JOB. IT IS A GOOD WAY TO WIND DOWN, TO SAVE TOOLS, AND IT HELPS MAKE FOR A GOOD START THE NEXT MORNING.

WHEN WINDING UP EXTENSION CORDS, ALTERNATE WINDINGS FROM DAY TO DAY OR WEEK TO WEEK. WIND WITH THE RIGHT HAND ONE DAY AND THE LEFT THE NEXT. IT IS PARTICULARLY IMPORTANT WITH RUBBER CORDS BECAUSE THEY WILL TWIST AND SPLIT. THE PLASTIC CORDS SIMPLY GET TWISTED AND TANGLED MAKING UNWINDING DIFFICULT.

TUESDAYS AND THURSDAYS...

MONDAYS, WEDNESDAYS, AND FRIDAYS.

SOME SAFETY TIPS

ACCIDENTS, BOTH PERSONAL AND TO THE TOOLS AND BUILD-ING, ARE ALWAYS PREVENTABLE. NEVER LEAVE A BLOCK OF WOOD ON THE GROUND WITH A NAIL THROUGH IT.

A GOOD REASON NOT TO WEAR SNEAKERS ON THE JOB.

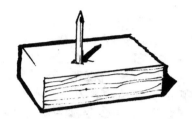

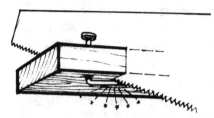

FOLD IT OVER, BUT BETTER STILL PULL IT. IT WON'T DULL A SAW WHEN YOU THINK YOU'RE CUTTING ALONG SIDE A NAIL.

NAILS ARE PARTICULARLY HAZARDOUS AT THE END OF LONG BOARDS. THEY CAN RIP OPEN A HAND OR LEG EVEN WHEN FOLDED OVER. NAILS STICKING OUT WHERE THERE IS PEOPLE TRAFFIC DOES A JOB ON CLOTHS AND SKIN. DOOR WAYS AND HALL WAYS ARE NO PLACES FOR PROJECTING NAILS.

KEEP WORK AREAS IN FRONT OF SAWS CLEAR. THROW CUT ENDS AWAY INTO A SCRAP PILE. STUMBLING OVER BLOCKS IN FRONT OF A SAW IS NOT ONLY DANGEROUS IT IS ALSO AGGRAVATING AND WHEN ONE IS AGGRAVATED ONE GETS CARELESS. KEEP DECKS CLEAR OF 2x4 BLOCKS, THEY ARE GREAT ANKLE TWISTERS.

TRY TO PREVENT WOOD BRACES FROM PROJECTING TOO FAR OUT AT ABOUT FOREHEAD HEIGHT. IT IS A HEIGHT THAT IS EASILY MISSED WITH THE EYE, BUT EASILY FOUND WITH THE HEAD.

IF A BOARD MUST BE THERE HANG A FLAG ON IT.

CLEAN UP FREQUENTLY; A CLEAN JOB IS A LESS ACCIDENT PRONE JOB.

DON'T USE TOOLS BEYOND THEIR CAPACITY. A SHOVEL IS NOT A PRY BAR.

3/4" SHIM OR TEN d. NAIL   BUTCHER'S TWIST KNOT

STRING AND SHIM SET UP FOR STRAIGHTENING WALLS. A SIMPLE LOOP TO START AT ONE END AND THE OLD FASHIONED "BUTCHERS TWIST" AT THE OTHER. THAT'S WHAT THE OLD TIME BUTCHERS USED TO CALL IT WHEN THEY TIED UP THEIR MEAT.

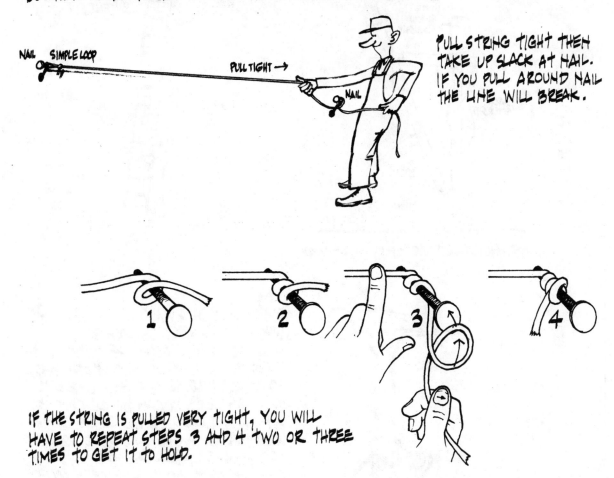

NAIL   SIMPLE LOOP   PULL TIGHT →   NAIL

PULL STRING TIGHT THEN TAKE UP SLACK AT NAIL. IF YOU PULL AROUND NAIL THE LINE WILL BREAK.

1   2   3   4

IF THE STRING IS PULLED VERY TIGHT, YOU WILL HAVE TO REPEAT STEPS 3 AND 4 TWO OR THREE TIMES TO GET IT TO HOLD.

MISCELLANEOUS TIPS

THE LADDER IS AN AWKWARD THING TO MOVE, ESPECIALLY A LONG EXTENSION. IT CAN BE MOVED QUITE EASILY THIS WAY, EVEN FULLY EXTENDED. IT REQUIRES SOME BALANCING WITH ARMS, BODY AND HEAD.

MOVING A HEAVY LADDER FULLY EXTENDED

KEEP FOOT OF LADDER CLOSE TO GROUND.

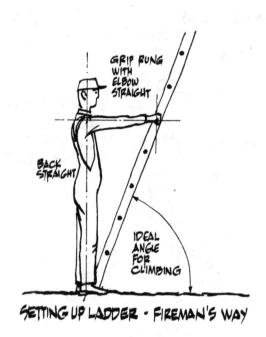

GRIP RUNG WITH ELBOW STRAIGHT

BACK STRAIGHT

IDEAL ANGLE FOR CLIMBING

SETTING UP LADDER - FIREMAN'S WAY

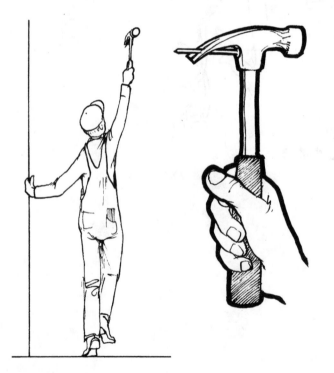

EVERY ONCE IN A WHILE THERE IS A TIME WHEN A NAIL MUST BE LOCATED TOO HIGH OR TOO FAR FROM THE WORK PLATFORM, TO BE REACHED WITH BOTH HANDS (ONE FOR HOLDING THE NAIL, ONE FOR THE HAMMER). IT CAN BE REACHED IF A NAIL IS TUCKED IN THE CLAWS OF THE HAMMER WITH THE NAIL HEAD AGAINST THE HAMMER NECK AND THE NAIL POINTING OUT. SOME HAMMERS ACCEPT A NAIL THIS WAY BETTER THAN OTHERS. WEDGE THE NAIL IN AND IT SHOULD STAY. THEN REACH UP AND START THE NAIL WITH A SWING. ONCE IT'S HELD FIRMLY BY THE WOOD, THE HAMMER IS REVERSED AND THE NAIL DRIVEN HOME.

THE THUMB HOLDS PENCIL FIRMLY AGAINST POINTER AND MIDDLE FINGERS. FOR LINES UP TO ONE INCH FROM THE EDGE, THE MIDDLE FINGER RIDES ALONG THIS EDGE.

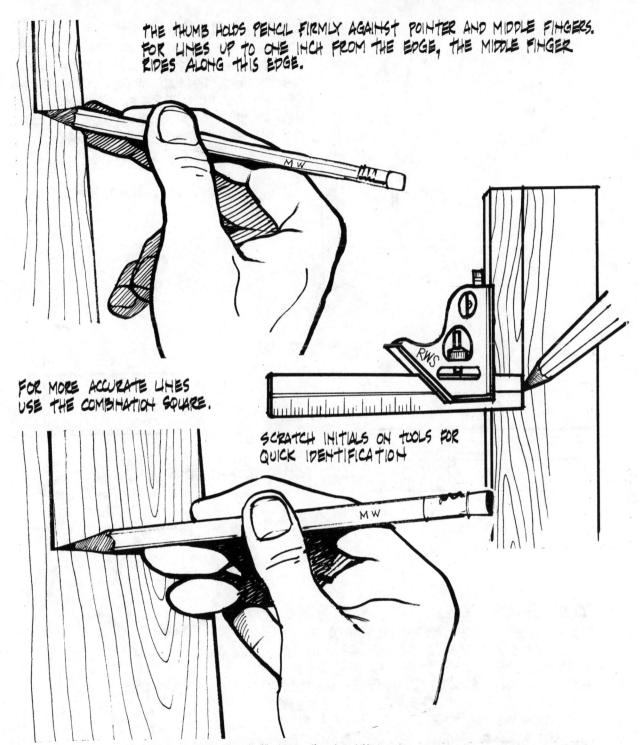

FOR MORE ACCURATE LINES USE THE COMBINATION SQUARE.

SCRATCH INITIALS ON TOOLS FOR QUICK IDENTIFICATION

LINES UP TO FOUR INCHES FROM THE EDGE CAN BE DRAWN USING THE SMALL FINGER AS THE GUIDE. WHATEVER FINGER IS USED, THEY MUST ALL BE KEPT RIGID BY BRACING THEM AGAINST EACH OTHER.

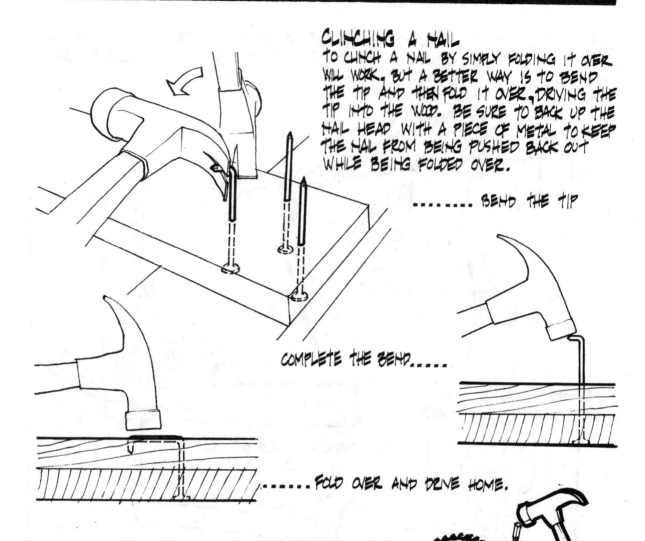

CLINCHING A NAIL

TO CLINCH A NAIL BY SIMPLY FOLDING IT OVER WILL WORK, BUT A BETTER WAY IS TO BEND THE TIP AND THEN FOLD IT OVER, DRIVING THE TIP INTO THE WOOD. BE SURE TO BACK UP THE NAIL HEAD WITH A PIECE OF METAL TO KEEP THE NAIL FROM BEING PUSHED BACK OUT WHILE BEING FOLDED OVER.

.... BEND THE TIP

COMPLETE THE BEND......

.... FOLD OVER AND DRIVE HOME.

ROUGH SHARPEN & SETTING SAW BLADE

EVEN THOUGH A GOOD CARPENTER HAS A GOOD SUPPLY OF SHARP BLADES ON THE JOB, THERE IS BOUND TO BE A TIME WHEN ONE HAS TO BE SHARPENED ON THE JOB. TO SHARPEN, SUPPORT THE BLADE, AS CLOSE TO THE TOOTH BEING DRESSED, AND FOLLOW THE EXISTING ANGLES WITH A FIRM FORWARD STROKE OF THE FILE. AFTER SHARPENING, LAY THE BLADE ON A PIECE OF PLYWOOD AND SET EVERY OTHER TOOTH WITH A NAIL SET AND HAMMER. TURN THE BLADE OVER AND REPEAT ON THAT SIDE. THE MORE SET THERE IS THE EASIER AND ROUGHER THE CUT.

THERE ARE MANY PROBLEMS THAT CROP UP ON A JOB THAT SEEM TO HAVE NO SOLUTIONS. OF COURSE SOLUTIONS ARE POSSIBLE, BUT YOU NEVER KNOW WHERE THOSE SOLUTIONS WILL COME FROM, SO BE RECEPTIVE TO IDEAS. ALSO BE AWARE OF OTHERS' PROBLEMS. YOU MAY HAVE THE SOLUTIONS THEY NEED.

MY DAD TOLD ME OF ONE SITUATION WITH A PROBLEM THAT WAS SOLVED IN JUST THAT WAY. HE WAS TRIMMING OUT A DORMER WITH CROWN MOULDINGS THAT MET AT THE RAKE AND EAVES. THE ANGLE OF THE CUT WAS IMPOSSIBLE TO FIGURE. A FELLOW CARPENTER BELOW LOOKED UP AND SAID "EYEBALL AND CUT IT!" SO MY DAD DID IT, AND TO HIS AMAZEMENT IT WORKED!

THEN MONTHS LATER, ON ANOTHER JOB, THIS SAME FELLOW CARPENTER WHO HAD BEEN BELOW GIVING THE ADVICE, WAS UP AGAINST THE SAME PROBLEM HIMSELF AND WAS UNABLE TO SOLVE IT. THIS TIME MY DAD WAS BELOW AND SAID TO HIM "EYEBALL AND CUT IT!" TO THE AMAZEMENT OF BOTH PARTIES IT WORKED AGAIN. YOU NEVER KNOW.

IN CONCENTRATING ON CARPENTRY TRICKS OF THE TRADE I HAVE NOT SHOWN OR DISCUSSED MANY SAFETY PRACTICES OR DEVICES, TAKING IT FOR GRANTED THAT BASIC PROCEEDURES WOULD BE PRACTICED BY THE READER. SAFETY GOGGLES, PROTECTIVE CLOTHING, SAFE LADDERS, TEMPORARY RAILINGS, AND MANY OTHER PRODUCTS AND PRACTICES ARE PART OF GOOD CARPENTRY WORK. OSHA RULES, MANUFACTURERS INSTRUCTIONS, BUILDING CODES, AND, MOST OF ALL, COMMON SENSE, MUST TAKE PRIORITY OVER HAMMERS AND NAILS IN SUCCESS- FUL CARPENTRY.

CUTTING FLASHING IS BEST DONE WITH A UTILITY KNIFE. A FRAMING SQUARE AGAINST A PARTITION PLATE THAT HAS BEEN MARKED OFF FOR THE FLASHING SIZE MAKES FOR SQUARE CUTS OF EQUAL LENGTHS.

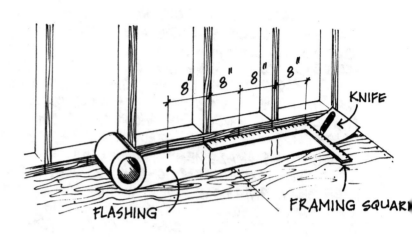

FLASHING

FRAMING SQUARE

ANY SIZE NAIL CAN BE STARTED WITH THIS ONE-HAND NAILING METHOD.

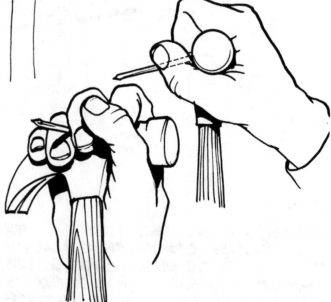

HOLD A NAIL BETWEEN THE FINGERS WITH THE NAIL HEAD AGAINST THE SIDE OF THE HAMMER HEAD. WITH ONE BRISK SWING, DRIVE THE POINT OF THE NAIL INTO THE WOOD.

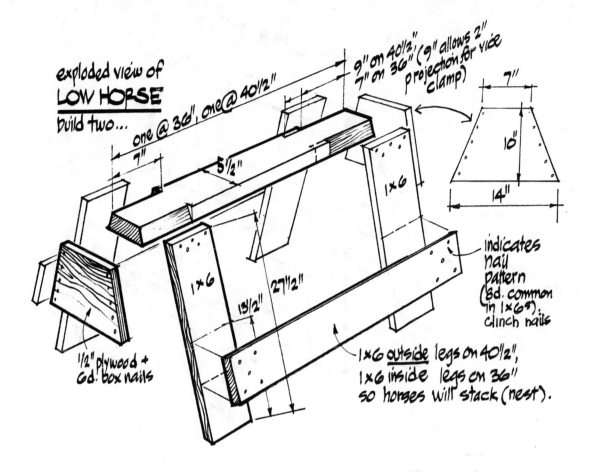

exploded view of **LOW HORSE** build two...

one @ 36", one@ 40½"

7"

5½"

9" on 40½"
7" on 36" (9" allows 2" projection for vice clamp)

7"

10"

14"

1×6

1×6

27½"

13½"

indicates nail pattern (8d. common in 1×6s). clinch nails

1×6 <u>outside</u> legs on 40½", 1×6 <u>inside</u> legs on 36" so horses will stack (nest).

½" plywood + 6d. box nails

CLINCH 6d OR 8d NAILS IN LEGS. SEE CLINCHING DETAILS.

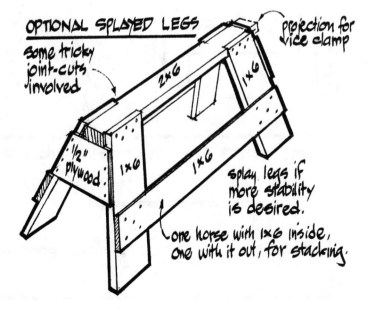

OPTIONAL SPLAYED LEGS

some tricky joint-cuts involved

projection for vice clamp

2×6

1×6

½" plywood

1×6

1×6

splay legs if more stability is desired.

one horse with 1×6 inside, one with it out, for stacking.

SAW HORSE

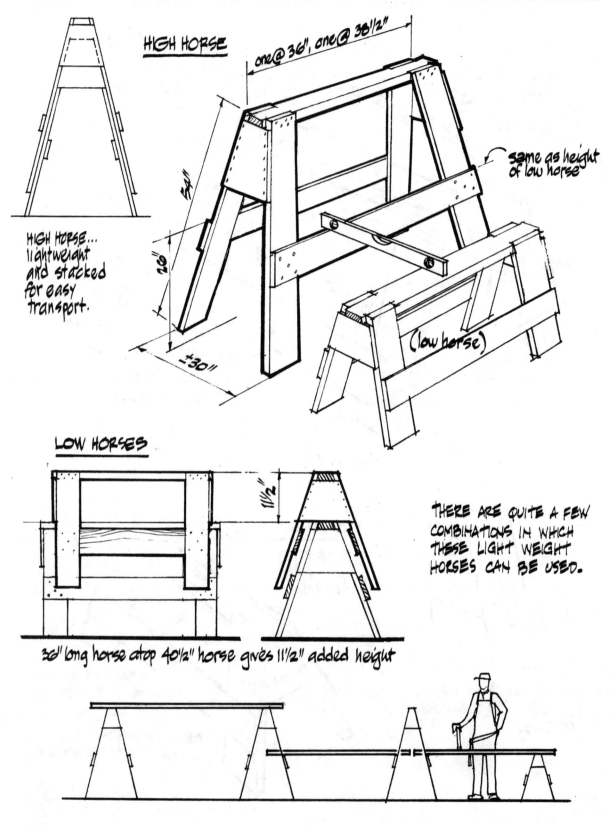

HIGH HORSE

one @ 36", one @ 38½"

same as height of low horse

54"

29"

± 30"

HIGH HORSE...
lightweight
and stacked
for easy
transport.

(low horse)

LOW HORSES

1½"

THERE ARE QUITE A FEW
COMBINATIONS IN WHICH
THESE LIGHT WEIGHT
HORSES CAN BE USED.

36" long horse atop 40½" horse gives 11½" added height

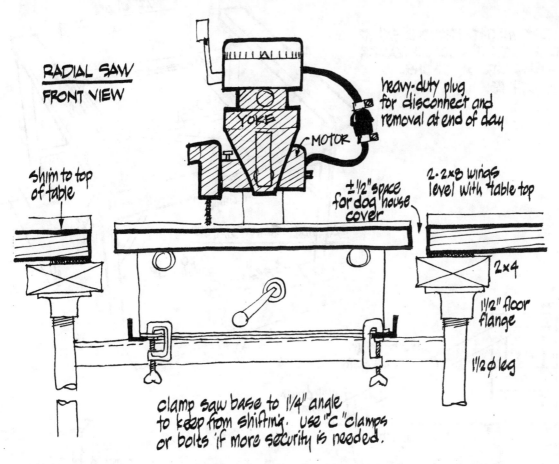

RADIAL SAW
FRONT VIEW

heavy-duty plug
for disconnect and
removal at end of day

YOKE

MOTOR

shim to top
of table

± 1/2" space
for dog house
cover

2-2x8 wings
level with table top

2x4

1 1/2" floor
flange

1 1/2 Ø leg

clamp saw base to 1 1/4" angle
to keep from shifting. Use "C" clamps
or bolts if more security is needed.

REMOVE YOKE WITH MOTOR AT END OF DAY UNLESS SECURITY IS NO PROBLEM. A HEAVY DUTY PLUG INSTALLED ON THE WIRE FROM THE MOTOR MAKES THIS EASY.

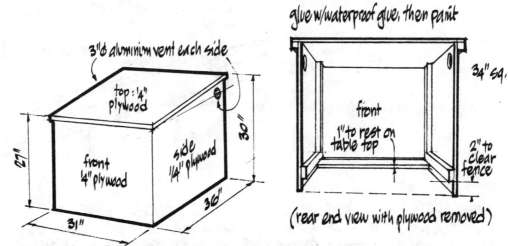

3"Ø aluminum vent each side

top: 1/4"
plywood

front
1/4" plywood

side
1/4" plywood

27"

30"

31"

36"

glue w/waterproof glue, then paint

3/4" sq.

front

1" to rest on
table top

2" to
clear
fence

(rear end view with plywood removed)

A DOG HOUSE COVER FOR THE RADIAL SAW KEEPS THE ELEMENTS OFF. A NAIL HERE AND THERE WILL HELP TO SECURE IT.

RADIAL SAW STAND

THIS LIGHT WEIGHT PORTABLE STAND
IS FOR A "ROCKWELL 10" CONTRACTOR
MODEL SAW. ADJUST DIMENSIONS
TO FIT OTHERS.

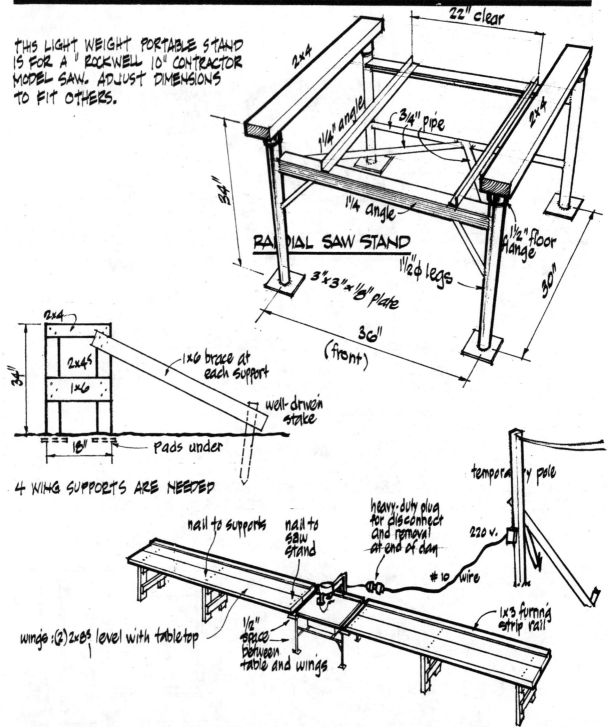

2x4

22" clear

1¼" angle

3/4" pipe

1¼ angle

RADIAL SAW STAND

2x4

34"

36"
(front)

1½" floor flange

1½"⌀ legs

3"x 3"x ⅛" plate

30"

2x4

34"

2x4's

1x6

18"

1x6 brace at
each support

well-driven
stake

Pads under

4 WING SUPPORTS ARE NEEDED

nail to supports

nail to
saw
stand

heavy-duty plug
for disconnect
and removal
at end of day

temporary pole

220 v.

10 wire

wings :(2)2x8's level with table top

½"
space
between
table and wings

1x3 furring
strip rail

AFTER THE STAND IS SET UP AND WINGS ARE ATTACHED, EYEBALLED PARALLEL TO THE
SAW'S TABLE FENCE, SNAP A PARALLEL CHALK LINE FROM END TO END. NAIL A 1x3 FURR-
ING STRIP, ON EACH WING, TO THIS LINE. BE SURE THERE IS A GAP BETWEEN WINGS AND
SAW'S TABLE TO ALLOW FOR DOG HOUSE COVER. MAKE UP A LENGTH (25' OR SO) OF
#10 WIRE, WITH A HEAVY DUTY PLUG FOR THE 220V CONNECTION TO WORK POLE.

108

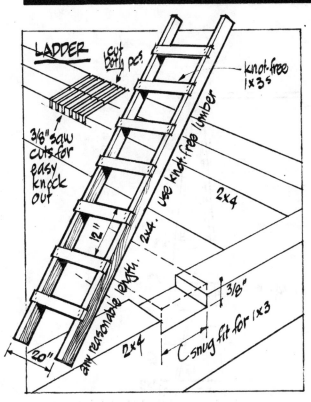

LADDER

cut both pcs.

knot-free 1×3s

3/8" saw cuts for easy knock out

use knot-free lumber

2×4

12"

2×4

any reasonable length.

3/8"

20"

2×4

snug fit for 1×3

ONE OF THE FIRST THINGS NEEDED ON A JOB IS A LADDER. USUALLY THESE LADDERS ARE 10' OR 12', BUT IF A LONGER ONE IS BUILT, USE STRAIGHT GRAINED FIR FOR GREATER STRENGTH. MARK AND CUT BOTH RAILS AT THE SAME TIME. IF YOU MAKE TWO, CUT ALL FOUR AT THE SAME TIME.

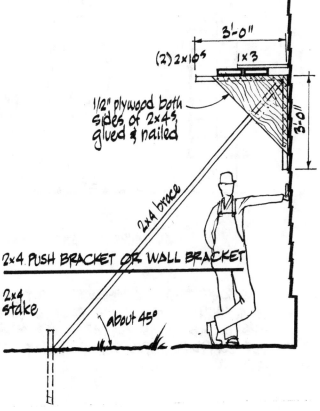

3'-0"

(2) 2×10s

1×3

1/2" plywood both sides of 2×4s, glued & nailed

3'-0"

2×4 brace

2×4 PUSH BRACKET OR WALL BRACKET

2×4 stake

about 45°

SLIP BRACKET OVER A LONG 2×4 AND LEAN IT AGAINST THE BUILD-ING. SET UP A MATCHING ONE AND THEN DRIVE A STAKE AT THE BASE OF EACH DIAGONAL 2×4 BRACE. THE BRACKETS ARE NOW READY FOR THE 2×10 PLANKS.

109

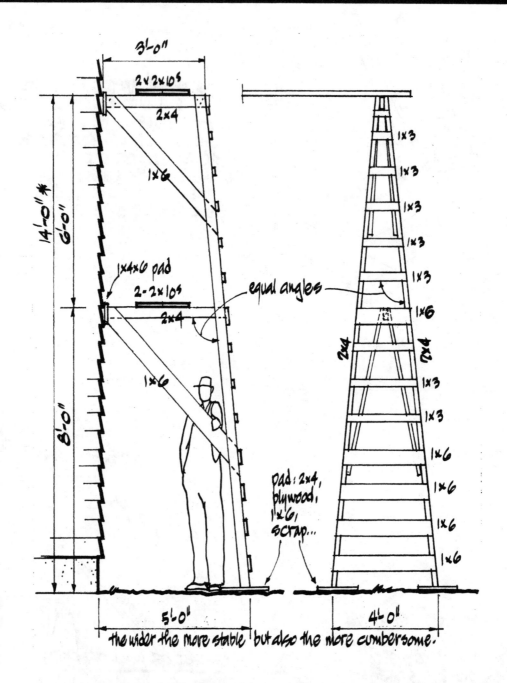

3'-0"

2x2x10s

2x4

1x6

14'-0"*

6'-0"

1x4x6 pad

2-2x10s

2x4

equal angles

1x6

8'-0"

pad: 2x4, plywood, 1'x6', scrap...

1x3
1x3
1x3
1x3
1x3
1x6
2x4
2x4
1x3
1x3
1x6
1x6
1x6
1x6

5'-0"

4'-0"

the wider the more stable, but also the more cumbersome.

THIS SCAFFOLDING IS CUMBERSOME, BUT EFFECTIVE. A GOOD VARIATION ON THIS IS TO BUILD JUST THE TOP PART 6' TO 8' HIGH. ALTHOUGH YOU CAN CLIMB UP THE OUTSIDE, IT IS BETTER TO USE A LADDER. ALWAYS PUT WOOD PADS UNDER SCAFFOLDING POSTS AND LEGS TO KEEP THEM FROM SINKING INTO THE GROUND.

TOOL BOXES SHOULD HAVE A PLACE FOR NAIL SETS, BLOCK PLANE, FRAMING SQUARE, COMBINATION SQUARE, CHISELS, SHARPENING STONE, OIL CAN, HATCHET, AND LEVEL. THE LEVEL SHOULD BE PROTECTED IN A SEPARATE COMPARTMENT. THE CHISELS AND HATCHET SHOULD HAVE THEIR EDGES PROTECTED. A SEPARATE BOX FOR HAND SAWS IS A MUST AND THE BLADES MUST BE SEPARATED FROM EACH OTHER. A THIRD CATCH-ALL BOX FOR TOOLS LIKE HAMMERS, NAIL PICKS, FLAT BARS, SCREW DRIVERS, AND WRENCHES IS HELPFUL. THIS BOX MUST BE STRONG ENOUGH TO HANDLE THE HEAVIER WEIGHT, AND NOT TOO DEEP (TO PREVENT PILING UP OF TOO MANY TOOLS).

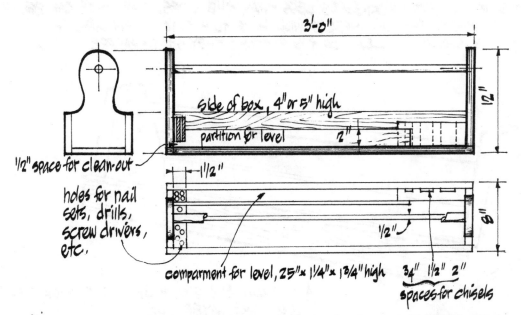

3'-0"

12"

side of box, 4" or 5" high

partition for level

2"

1/2" space for clean-out

holes for nail sets, drills, screw drivers, etc.

1 1/2"

1/2"

8"

comparment for level, 25" x 1 1/4" x 1 3/4" high

3/4" 1 1/2" 2" spaces for chisels

THERE ARE MANY THINGS ONE CAN DO TO KEEP TOOLS FROM RATTLING AROUND.

TOOL BOX

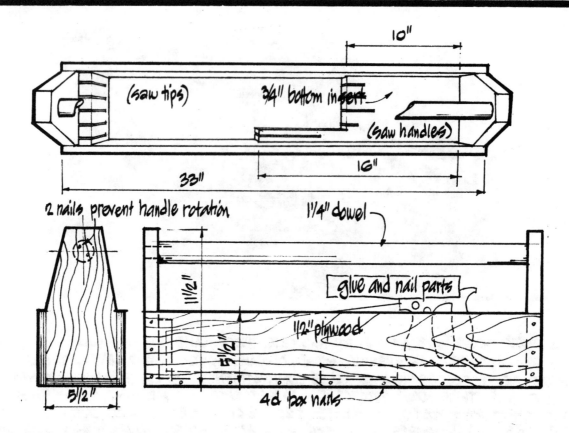

10"

(saw tips)   3/4" bottom insert   (saw handles)

33"   16"

2 nails prevent handle rotation   1 1/4" dowel

11 1/2"   glue and nail parts

1/2" plywood

5 1/2"

5 1/2"   4d box nails

RIP ALL BOTTOMS, ENDS, AND INSERTS 5 1/2" WIDE. THEN RIP ALL SIDES 5" WIDE, AND PARTITIONS AS REQUIRED. SIDES AND BOTTOMS ARE ALL THE SAME LENGTH. THE HANDLE IS TWO END THICKNESSES LESS THAN THE SIDES. THE ENDS CAN BE SIMPLE OR FANCY. THE SAW INSERTS WILL HAVE TO FIT YOUR SAW LENGTHS. THE BLADES SHOULD BE CLEAR OF THE BOTTOM; THEY STAY SHARPER THAT WAY.

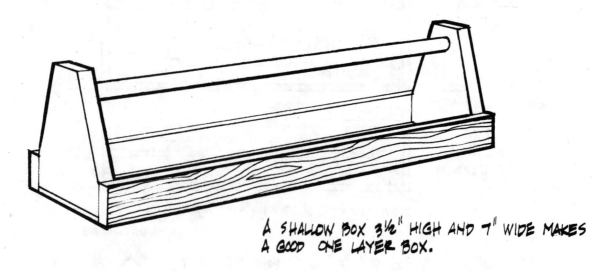

A SHALLOW BOX 3 1/2" HIGH AND 7" WIDE MAKES A GOOD ONE LAYER BOX.